THE BLACK DEATH IN LONDON

The Great Plague Epidemic and its impact on the population through social and cultural changes

by

Jack Stew Barretta

TABLE OF CONTENTS

Introduction

In 1348, the first major epidemic since the Plague of Justinian swept across virtually all of Europe dramatically changing its disease climate. This plague not only altered the biological conditions under which Europeans lived but by killing as much as half the population during the initial epidemic wave, it dramatically affected social conditions. Although the effects of the plague were felt across the entire European continent and beyond, the social changes the plague produced across the various kingdoms of Europe varied greatly because of local cultural conditions. In England, the epidemic and the decline in population facilitated social and cultural changes. Changes in material culture affected relationships between rats, fleas, and people, and these changes help to explain the eventual disappearance of the plague from Britain.

There was a gradual increase in the general level of affluence during the 300 years plague was endemic in England, although during this period there were also increasing disparities between the mortality rates of poor people and those who were better off. Almost certainly, these changes involved differences in human behavior and material culture. An exploration of the conditions of the plague and its impact on the early modern period requires an investigation of conditions during the earlier part of the epidemic. Comparing information from the duration of the epidemic may explain not only why the poor suffered greater rates of mortality than the rich, but also how a single pathogen, *Yersinia pestis*, presented so differently throughout the various pandemics.

In this manuscript, I examine in detail the patterns of plague outbreaks in Britain and affirm that plague, the disease produced by *Yersinia pestis,* was the disease of greatest significance during the late medieval and early modern periods. Additionally, I examine how, because of the unique way in which plague is spread, changes in material wealth and living patterns affected the course of epidemics. Some of the strategies used to combat plague were useless or worse, and yet during the more than 300 years that plague was endemic, 1348 to 1665, changes in lifestyles created an environment in Britain that was much less conducive for the survival of *Y. pestis*. In the years after plague disappeared from Britain, social and material changes continued, so that when the plague was reintroduced to Britain in the early 1900s, it made few inroads into the country and produced only a small outbreak. I examine the contemporary societal understanding of epidemics as portrayed by public health measures and common treatment modalities as a way to interpret the changes in epidemic patterns that occurred in this still enigmatic disease.

Plague affected personal and societal behaviors and so it is important to investigate how the changes in human behavior and social patterns affected the pattern of disease spread. The pattern of plague epidemics among humans is complex, involving as it does the complex interaction of the relationship between fleas and rats to humans. Thus, anything that people did that changed this relationship had an impact on the extent and severity of plague outbreaks. Human behavior has affected, negatively and positively, mortality levels inflicted throughout plague epidemics. Failure to provide palliative care due to fear increased mortality levels, while human behaviors that changed environmental conditions or produced loud noise and commotion that could have frightened rats away might have decreased morbidity rates. Blacksmiths, whose work environment necessarily involves noise and does not provide anything edible for rats, had lower than average mortality rates during plague epidemics.

Although plague posed a danger throughout Britain, epidemics in small towns and villages were sporadic, while the plague was virtually ubiquitous in London from the time it first appeared in 1348 until 1665 when the last major epidemic occurred there. London was the country's largest city, a major shipping port, the economic hub of England, the city most closely associated with the seat of government, and the center of plague in England. Because of this, plague deaths were watched and tracked more closely in London than in the rest of Britain. In 1519, causes of death began to be recorded sporadically during plague years, but by 1629 the Bills of Mortality recorded the causes for all deaths in London. These records show that there were only a very few years when the plague was completely absent, and equally significantly, records indicate that plague becomes epidemic only sporadically. By the late 1500s, people had begun to notice that in England plague was more devastating among the poor than among the wealthier.

Europeans have been studying the series of plague epidemics that ravaged Europe between 1347 and the 1720s since the plague first erupted across the continent in 1347. Some Europeans were aware of the disease's outbreak in the East and had followed its progress before it entered Europe. The *Historia Roffensis*, a chronicle of the Rochester cathedral priory, describes the beginning of plague: "A great mortality of men began in India and, raging through the whole of infidel Syria and Egypt and also through Greece, Italy, Provence, and France arrived in England, where the same mortality destroyed more than a third of the men, women, and children." Guy de Chauliac, the surgeon to Pope Clement VI, noted its spread throughout the East and into the West. In

Great Surgery written in about 1363, Chauliac wrote:

> *I call it great, because it covered the whole world, or lacked little of doing so. For it began in the East, and thus casting its darts against the world, passed through our region toward the West. It was so great that it left scarcely a fourth part of the people. And I say that it was such that it's like has never been heard tell of before; of the pestilences in the past that we read of, none was so great as this. For those covered only one region, this the whole world; those could be treated in some way, this is none... Many were in doubt about the cause of this great mortality. In some places, they thought the Jews had poisoned the world: and so they killed them. In others, that it was the poor and deformed: and they drove them out. In others, that it*

was the nobles: and they feared to go abroad.

Treatises by learned physicians admitted that some sort of natural phenomenon or process might be responsible for the disease's prevalence, although most believed that God was ultimately responsible. Europeans were unsure if the disease was a secondary result of a natural phenomenon, or if God's primary intention was to punish them for their lack of faith as much as to punish infidels for adhering to the wrong faith.

When the plague arrived in Europe, it became a focus of inquiry. A dozen plague treatises written by physicians and surgeons during the first two years of the epidemic have survived. In these treatises, the authors discuss methods for avoiding plague and treating victims. Also, the authors attempted to explain the causes of the disease. Authors of these treatises explored many possible general causes of the plague, the majority of which included the production of poisonous vapors. Various explanations for the production of these vapors were offered, including planetary conjunction on March 20, 1345, a battle between solar rays and the Indian Ocean, and earthquakes in 1347. Medieval physicians and surgeons assumed that if they could discover the causes of the disease, they would be able to help prevent outbreaks and cure victims. Despite years of study, many questions about the plague outbreaks remain unanswered, notably, why did plague arrive when it did and what caused it to disappear after several hundred years of devastation? Although it has long been accepted that closing houses with plague victims and their families locked inside increased European plague mortality rates, the full impact of human behavior on plague epidemics is yet to be discovered.

Not only is an overall explanation for the pattern of plague pandemics still a mystery, but even the general pattern of plague epidemics remains unresolved, in large part because it is unclear to what extent Europeans distinguished between epidemics of different diseases.

Interest in questions and problems produced by plague epidemics of the past has waxed and waned. During the centuries when plague ravaged Europe, it was a focus of interest and inquiry; in the years from 1348 to 1500 as many as 1,000 plague treatises spread throughout Europe. However, because very little progress was made in treating or preventing plague when other diseases entered Europe, the focus of inquiry shifted away from plague to new diseases such as syphilis. Nonetheless, while plague remained a common but terrifying feature of the disease landscape, medical treatises providing methods for avoiding and treating plague continued to be written and printed. During the Enlightenment, there was great faith that human study and ingenuity would almost inevitably produce progress in human health and living conditions. As a result, interest in the chaos caused by past scourges subsided. Plague outbreaks slowly receded east with the last major outbreak in London in 1665, in Marseilles in 1720, and the final outbreak in Moscow in 1771. Smallpox became controllable when Lady Mary Wortley Montagu (1689-1762) introduced to England a method of inoculation she had learned of in Turkey; thus, the investigation of plague, which was seen to be a symptom of past failures, fell out of favor. Interest in plague did resurface in London in the 1720s as a practical response to the threat of a plague outbreak that was raised by the epidemic in Marseille, and then again in the 1750s when England feared an outbreak because of plague in the Crimea. For the most part, the influence of the Enlightenment made future improvements more significant than the investigations of large-scale destruction and societal failures of the past.

However, by the beginning of the twentieth century, events conspired to make the study of plague and the devastation it had wrought in previous centuries appear of increasing relevance. These events called attention to the devastating social impact of the plague as much as to its medical and health aspects. In 1894, bubonic plague erupted in Hong Kong and began to spread around the world. This outbreak drew attention to the plague but, as importantly, the large-scale disruption and chaos caused by World War I produced an increased interest in the extensive disruptions produced by the initial plague outbreak of 1348 in Europe. Although World War I was followed by a devastating global flu outbreak that killed "millions upon millions of people in a year or less," these deaths caused by the 1918 epidemic do not seem to have spurred inquiry into historic disease outbreaks. It was World War I, which provided a firsthand demonstration of worldwide chaos, together with the Modern Pandemic, a plague pandemic that began in Hong Kong in 1894, that eventually increased interest in earlier plague pandemics and the devastation and societal changes they produced.

Plague continues to be of interest because, from the vantage point of the first decade of the twenty-first century, the threat of inescapable pandemics is again all too real. The threats seem to be myriad, as new diseases are constantly being discovered and old diseases are being found in new places. AIDS continues to spread and kill worldwide, and although the virus that causes it has been isolated, the speed with which it mutates has not only confounded attempts to treat the infected and to create a vaccination, it has also challenged conceptions about how viruses behave. The Second Pandemic and HIV epidemics occurred centuries apart and epidemiologically behave very differently, yet the problems and questions that societies encounter as they attempt to control the spread of a fatal disease are remarkably similar. The study of plague epidemics of medieval and early modern Europe can serve as a warning of possible dangers and provide an encouraging example of a society that managed to survive, despite devastating death tolls.

Although HIV is a widespread infectious disease threatening the world in the twenty-first century, it is by no means the only infection that is challenging societies and epidemiologists. The list of new diseases and the potential for new diseases seem almost endless. Ebola, a hemorrhagic fever, has been breaking out sporadically in isolated communities in west Central Africa. In addition to humans, it has ravaged animals such as gorillas, and only after extensive searching for the host population have three species of fruit bats been identified as potential reservoirs. Moreover, the recognition of Hantavirus pulmonary syndrome (HPS) in 1993 drew attention to the fact that disease agents with the potential to produce new epidemics are still a threat despite modern science. HPS also served as a reminder that traditional learning and social patterns often develop in response to disease threats. The Navajo people of the southwest have traditional routines of cleanliness that were based on mythological events. These beliefs discouraged waste and garbage in the house, factors that might entice deer mice into having close contact with humans; thus, these beliefs and their associated behavior likely protected Navajo people from infection from the Hantavirus.

The Severe Acute Respiratory Syndrome (SARS) outbreak in Asia and the ongoing examples of avian flu jumping the boundary between species, from fowl to humans, have reminded people that despite all the medical knowledge and technology available in the twenty-first century, another wide-scale and the un-treatable human epidemic is a very real possibility. Societies might one day face an epidemic with morbidity and mortality rates similar to those of plague epidemics of the past; thus, an inquiry into these epidemics is not driven exclusively by a desire to examine the chaos of historic periods. The inquiry is also driven by a desire to understand the infrastructure and cultural strengths that permitted society to function despite mortality rates that sporadically exceeded 20 percent. These mortality rates were high enough to make even the task of burying victims an almost insuperable burden. In the light of the distress produced by the insignificant anthrax threats that occurred in 2001, the ability of historical societies to continue to function during plague epidemics, despite catastrophic mortalities, is truly awe-inspiring.

The study of the Second Pandemic provides an example of human survival and interaction with a mysterious and misunderstood epidemic force. Furthermore, a comparison of the plague's impact during the initial epidemics with the impact of the final epidemics of the pandemic serves as a reminder that human behavior can have a direct influence on disease patterns as well as the reverse. The study of plague epidemics of the past provides a reminder that broad societal factors can be as significant as medical care in determining the outcome of epidemics.

Studies of individual resistance to infection by HIV have provided evidence that plagues of the past have had a direct and significant influence on the genetic composition of Western Europeans. A genetic mutation that probably originated in Scandinavia and was distributed through Europe by Viking invaders appears to be the best known and most effective natural defense against AIDS. This mutation, known as CCR5-32, is believed to have originated, or at least to have been very rare, 2000 years ago, but about 700 years ago Europeans experienced a disease or other natural force that favored the survival of people who had this mutation; thus the mutation became much more common. At present, it is unclear what disease selected for this mutation, but it must have been a disease that devastated Europe in the late middle ages. Two diseases being investigated for providing this selective force are *Y. pestis* and the variola virus, which produces smallpox. People who have inherited this mutation from both parents appear to be immune to HIV, and HIV seems to develop more slowly in those who have inherited the mutation from only one parent. In addition to slowing the disease progress, having one copy of the mutated gene seems to confer some resistance against contracting AIDS. If it can be made, the identification of the disease or diseases selected for this genetic mutation in northern Europe will provide a powerful argument for that disease has been a significant factor in the epidemics of the medieval and early modern periods.

As society again faces the threats of large-scale epidemics, without medical cures, the study of the way people understood and responded to the Second Pandemic is once again relevant. I explore this pandemic by focusing on London because it played a pivotal role in the plague cycle and because information on its aggregate population is more complete, over more years than for other locations in England.

However, London also presents some difficulties for analysis; the patterns plague exhibited there were not always duplicated throughout the country. It had an atypical population just by being the largest city in Britain, and it had a relatively young, mobile, and male-dominated population. During the late medieval and into the early modern periods, London's population grew, even though its death rate often exceeded its birthrate, because people, both peasants, and merchants, were drawn to London for the economic opportunities it provided. Because of this in-migration, scholars have found that the poor, illiterate people of London are less easily tracked as individuals and members of specific family groups than are peasants. In the countryside some manorial records allow historians to track changes in the tenancy of individual freeholds, thus providing considerable information on individuals and their families. Despite the limitations, a concentration on London has many advantages. Although details from the early epidemics are scarce, London provides the researcher with increasingly complete records from the later epidemics. London records include several first-person accounts of London plague epidemics, and plague pamphlets, as well as parish and city records.

Further, because the plague was endemic in London throughout plague epidemics between 1348 and 1665 known as the Second Pandemic, residents of the city were under more pressure to modify their behaviors than were inhabitants of places where the plague was less common. The special place that London held about Britain as a whole and concerning plague epidemics has long been recognized by officials and scholars. Recognition of London's role in spreading plague throughout England can be seen in the plague orders, that were issued in 1625 by the King and his Privy Council, rather than by London authorities. The Royal orders were theoretically directed at the region beyond the London liberties, but they were targeted at London. These orders required the suppression of "the grievous Infection of the Plague, and to prevent the increase thereof, within the City of London and parts about it." Despite their lack of authority over London liberties, the King and his Privy Council clearly understood that controlling plague in London was important to controlling plague in the rest of the country.

Printed plague orders were variously issued both by the monarch in conjunction with the Privy Council and by the London Corporation. They spelled out the measures that were to be undertaken by individual citizens and by local government to limit the impact of the plague and to control its spread.

They were issued, reissued verbatim, and in edited versions. Surviving plague orders that date from between 1578 and 1665 provide valuable insights into how the plague was understood and how its transmission was explained in Stuart and Tudor England. Because so much of this material was repeatedly issued with very few changes, I think that the understanding and concepts demonstrated in these orders substantially predate the earliest of the printed versions. These official plague orders were often published together with medical advice from "the best learned in Physicke within this Realme."

In addition to the medical advice published along with the plague tracts, I have examined Simon Kellwaye's medical treatise *A Defensative Against the Plague*. It was published in 1593, during a plague epidemic, shortly after the first of the English plague orders were published. The work is dedicated to "Robert Devorax, Earl of Essex and Ewe, Viscount of Hereford, Lord Ferrer of Chartley, Borcher, Louayne, Master of the Queen's Horse, Knight of the noble order of the Garter." In the flowery dedication, Kellwaye says that he has written this advice in emulation of the heathen "Stoikes" and "Romanes" who sacrificed for their countries unlike people "in this declining dotage of the world [when] the most of men are prone to follow their preferments, delighting in self-love and greedy snatching at the top of fickle fortunes wheel." Kellaway says that he has written the treatise so that anyone who reads it can learn how to avoid the plague, "as also how to order, govern and cure those that are infected therewith." The title page further emphasizes that his medical advice is provided as a service to the country. The recommendations that Kellwaye offers are similar to those offered with the plague orders, but his treatise is longer and provides a much broader range of recipes and his tone is more florid. Extensive summaries of these orders, as well as the long medical treatise, was written by Simon Kellwaye, are included as appendices.

Nomenclature

The disease that devastated England and much of the rest of Europe for several hundred years has been referred to by many different terms. The initial wave of death that swept across Europe was often simply referred to as "the great mortality." The following couple of outbreaks were named individually; the second outbreak was referred to as the *pestis secunda* or *pestis puerorum* because the victims were predominantly young people. The next major outbreak was referred to as the *pestis tertia*, but after this outbreak in 1369, subsequent outbreaks were often referred to simply as "pestilence" or "the peste" or "sickness." Shrewsbury notes that in the 1500s the term almost certain to indicate plague was "the great sickness." The term Black Death did not come into common usage until after 1665.

Probably "Black Death" originated as an over-literal translation into English or a Scandinavian language of the Latin terms *pestis mors* or *mors alta*. These Latin terms had been intended to signify a terrible pestilence rather than the more literal Black

Death. More recently it has been commonly assumed that the term describes the color of a body infected by the plague. The term Black Death has also been linked to clinical symptoms of the disease. It has been widely assumed to indicate the occurrence of severe petechial hemorrhaging, or bruising, which produces black splotches all over the victim. Although plague, like many other diseases, can occasionally produce black splotches, there is no evidence that the term Black Death is in any way descriptive of plague symptoms.

The use of the term Black Death poses several difficulties. Although it did not originate until late in the 300-year plague epidemic cycle, the term Black Death was then used primarily to designate the initial 1348 outbreak in the medieval/early modern plague cycle and to differentiate the initial outbreak from the Great London Plague of 1665. These terms were intended to differentiate two great epidemics, not the disease per se. Over time, the final London epidemic has receded in significance and the term Black Death has come to signify the disease rather than a specific plague epidemic. This shift in meaning presents a problem because the term Black Death is not used about plague epidemics of the present, nor to the pandemic referred to as the Plague of Justinian that began in 540 A.D., although these epidemics are generally thought to have been the result of the same disease. A second problem created by referring to the first wave of the pandemic by the term Black Death and later epidemics by other terms is that it creates a false dichotomy between epidemics. Using the term Black Death to designate the initial medieval outbreak also creates the impression that the initial outbreak was of primary significance. The initial onslaught in 1348 was the most widespread and devastating; however, the full impact of the plague on European society and material culture was produced by the almost continuous presence of plague for over three centuries.

Because of all the misconceptions associated with the term Black Death, I use the term plague in this manuscript to refer to the epidemic disease that ravaged Great Britain from 1348 through 1665. The term *plaga*, Latin for stroke, came to be associated with these epidemics early in the plague cycle. It was used first by King Magnus Eriksson of Norway and Sweden in 1349, and by the fifteenth century, transformed into the English form of the word, the plague had come into common usage in England. I chose this term in preference to other options because it is both specific enough to be acceptable to modern sensibilities and general enough to allow for some ambiguity; it is used to describe any number of diseases and other events that produce epidemics or other disasters. The term plague serves as a reminder that the people afflicted by these epidemics did not know exactly what they were dealing with, even though they do seem to have considered plaguing a discrete and uniquely terrible malady.

Biological Overview: *Yersinia Pestis*

The disease that swept through Europe in 1347-1348 and then remained endemic for more than 300 years was caused by a vector-borne bacterium, *Yersinia pestis*. The nature of the disease and how it is spread explain, in large part, why medieval and early modern Europeans had so much difficulty understanding plague epidemics and controlling the spread of the plague.

In 1894 *Y. pestis* was first identified as the causative agent of the plague outbreak that was raging in Hong Kong. This discovery was made only a couple of decades after a disease was first conclusively linked to a specific microbe. Because of the life cycle of *Y. pestis* and the etiology of the disease it produces are both complex, Western Europeans do not seem to have associated plague with either rats or fleas until 1894, after *Y. pestis* was isolated in patients in plague hospitals in Hong Kong.

Subsequently, it was identified in dead rats that had been noticed in conjunction with plague epidemics in Yunnan and Hong Kong. The fact that dead rats were found in conjunction with human plague outbreaks did not provide incontrovertible evidence that rats, let alone fleas, were implicated in the human plague, but by 1905 both rats and one of their fleas, *Xenopsylla cheopis*, were linked to the spread of the plague. The Indian Plague commission included an entomologist and the commission was actively pursuing the possibility that fleas were instrumental in the spread of the plague. In 1908, work in India confirmed that fleas were instrumental in spreading plague from rats to people.

While plague can be spread in several ways, the primary means is by flea bites. After fleas bite an infected animal, they take in and then carry the blood and the bacteria to another animal or human. However, even this primary step in the infective process is not simple. Although all fleas can ingest bacteria along with blood if there is a sufficiently high level of *Y. pestis* in the blood, only in flea species with a proventriculus, a part of the foregut, is the infectious process likely to be completed.

Xenopsylla cheopis, a rat flea, is the most

efficient plague vector. In these rat fleas, the bacilli sometimes become established in the fleas' proventriculus. Because virulent bacteria can divide more rapidly than they can be excreted, the bacteria can form a solid mass that blocks the stomach of the flea. This creates what is referred to as a *blocked flea*. This blockage prevents the flea from absorbing food – blood – into its stomach when it eats. In this condition, the flea becomes hungry and, in an attempt, to satiate its hunger it feeds voraciously. In this process, the gullet becomes distended and overextended because the flea's stomach is blocked. When the flea takes in more blood than it can hold, blood is regurgitated into the next bite, and in this process, *Y. pestis* can be injected into a new host.

Flea species differ not only in their ability to become blocked and thus to spread plague but also in their living and nesting habitats, which also affect the fleas' likelihood of spreading plague. In general, fleas can be divided into two groups based on their patterns of living: those that live on their host and those that live in some sort of nesting material from which they can easily reach their hosts. The human flea, *Pulex irritans*, resides primarily in nesting material such as sheets and bedding and is very sensitive to light. *X. cheopis*, considered the most effective flea for spreading plague, on the other hand, primarily lives on its host and is light tolerant. Further *X. cheopis* can survive by eating organic material such as grain, which means that it could be successfully transported in shipments even when no rats were present.

Infection by *Yersinia pestis* produces septicemia, which is the presence of pathogenic bacteria in the blood, and it is this process that efficiently spreads the infection. Because flea feces can contain bacilli, it is theoretically possible that flea feces could spread infection; however, feces are not likely an effective means of spreading plague. *Y. pestis* bacteria that are passed through the flea's alimentary canal are less virulent than bacteria that have not done so and are too large to easily cross the skin barrier. Flea feces do contain enough bacteria to cause infection through skin abrasions, and flea feces may be responsible for occasional plague cases. If plague was occasionally spread by flea feces, it could explain infrequent variations in the disease's virulence. Nonetheless, the most common method by which plague is spread, by blocked fleas, ensures that only the most virulent *Y. pestis* survive because only the most virulent bacilli reproduce quickly enough to result in a blocked flea. This ensures that *Y. pestis* remains consistently virulent, unlike other vector-borne diseases like the rickettsioses bacteria that produce typhus, which can vary greatly in virulence.

This difference helps explain why the last epidemic explosions of plague in London in 1665 and Marseilles in 1720 produced exceptionally high mortality figures.

Although contemporary *Y. pestis* outbreaks differ from those of the Second Pandemic, the three plague forms known today appear to have been present. The three forms are bubonic, pneumonic, and septicemic; all three are produced by the same bacillus, *Y. pestis*. The bubonic form is considered the most common form in plague outbreaks of the current pandemic. The disease's common name, bubonic plague, takes its name from the buboes, or swellings, that the disease produces in the lymphatic glands.

Swelling is most severe in the gland nearest the site of infection, in most cases a bite by an infected flea. Fatal cases of the bubonic form may include abscesses within the lungs. These infections are known as secondary pneumonic plague. When these lung abscesses rupture, they allow the bacilli to be coughed up and spread to new victims who are said to have the primary pneumonic plague. Because the pneumonic form develops when bacilli are inhaled, the bubonic form periodically transforms into the much more fatal and potentially contagious pneumonic form. The septicemic form develops when the bacilli multiply rapidly in the blood instead of collecting in the lymphatic glands. Death can occur within hours of the first symptom.

Recent studies into the DNA of *Y. pestis* indicates that it is a recently emerged clone of *Yersinia pseudotuberculosis*, which in most cases produces only a mild illness in humans. *Y. pestis* is so similar to *Y. pseudotuberculosis* that, except for the historical significance of *Y. pestis* and its impact on human populations, the names could be changed to reflect that they are one species. Based on mutation chronology, *Y. pestis* is considered by geneticists to be between 1,050 and 20,000 years old. Historic records from the Plague of Justinian (A.D. 542) indicate that it is at least 1,500 years old, but even at the oldest postulated age of 20,000 years, it is a young disease. The relative youth of *Y. pestis* may explain the disease's seemingly erratic mortality rates. It is possible that the infectious level and the mortality rate have not yet reached an evolutionary balance; the disease has not "evolved to an optimum level of virulence." While it is especially true that in humans *Y. pestis* has not achieved this optimum level of virulence, even in rats, its virulence seems to fluctuate wildly.

The cycle of human infection by *Yersinia pestis* is complex. Though *Y. pestis* has played a significant role in human history, it is primarily a disease of rodents, and humans are only incidental to the bacillus' life cycle. It cannot be emphasized too much that human epidemics occur as a secondary result of an epizootic, an epidemic within a rodent population. Epidemics only occur when humans are living near fleas and to animals more resistant to the disease, in other words, to animals that are capable of surviving an outbreak better than humans do. Thus, any changes that affect the relationships among rats, fleas, and humans can alter human morbidity rates.

Occasional human cases can occur due to accidental contact with fleas from an infected animal, but a full-scale epidemic almost by definition requires the full-scale die-off of rodents in close contact with humans. *Y. pestis* is endemic within some wild animal populations, typically ground-dwelling rodents, in what is called a sylvatic phase. There are now several permanent reservoirs, inveterate focus areas, where plague exists in a sylvatic phase, spread widely around the world, on all continents except Australia and Europe. Prairie dogs and ground squirrels in the Southwestern U.S. currently are carriers of *Y. pestis*. Other, longer-term inveterate foci include Central Asia, Siberia, the Yunnan region, Iran, the Kurdistan plains around the Caspian Sea, Libya, Arabia, and East Africa. Western Europe now has no sylvatic *Y. pestis* focus. During the plague's 300-plus-year infestation of Europe, however, there must have been areas where the plague was endemic among rodents, although, it is probable that the primary animals involved in Western Europe were rats cohabiting with humans rather than wild rodents. It seems clear that the plague was enzootic among London rats and that some plague outbreaks were caused by epizootics amongst the rats rather than by reintroduction of additional or new *Y. pestis*.

Because environmental conditions underground (e.g., cool and damp) favor the survival of both fleas and *Y. pestis*, fairly resistant burrowing animals that live in large communities are optimum carriers. To maintain the plague as an endemic disease, its host animals must be fairly resistant to *Y. pestis*, although even resistant species or individuals can be infected if injected with high enough doses of *Y. pestis*. Under certain environmental conditions, the infection becomes epizootic. Chester Rail has postulated that basic nutrients such as selenium and iron may play a part in this cycle and thus can affect both the virulence and the human or animal resistance. It has also been suggested that iron levels may have played a significant role in varying mortality rates among humans and have been especially important in mortality variations between men and women. However, it is not known whether the expected higher level of anemia in women would have been enough to provide them protection. Variations in mortality figures seem erratic and are more likely to have been affected by exposure levels.

As animals become sick and die, their fleas are driven to seek new hosts. Both fleas and rats (or possibly some other susceptible rodents living close to humans) are necessary for an epidemic. Because rats are so susceptible to *Y. pestis*, several conditions must be met for a protracted and serious outbreak of plague among humans. There must be an enormous rat population, or a reservoir population of some species of wild burrowing animals that occasionally comes in contact with rats or humans, or frequent infusions of plague bacilli. In England, there is no evidence that wild animals played a significant role in either spreading the plague or being a reservoir between human outbreaks. Instead, it seems likely that the rats of London served as the plague reservoir. Humans themselves cannot serve as a reservoir for the disease because they are simply too vulnerable. Even in cases of extreme septicemic infection in humans, it is unlikely there are enough bacilli in human blood to consistently infect fleas. In the pneumonic form, human-to-human contagion occurs, but this form occurs almost exclusively in conjunction with the bubonic form. Outside of a host, *Y. pestis* bacilli survive better in a cool, damp environment than they do in a hot, dry one, so the pneumonic form of plague is more easily spread in a cool climate.

Therefore, the pneumonic plague was probably more prevalent in Europe than in places such as India, Egypt, and South Africa, where the plague has been closely observed in modern outbreaks. Additionally, *Y pestis'* temperature preferences suggest that the pneumonic form of the plague was a significant factor in plague epidemics of late fall and winter. The initial symptoms for these two plague forms are quite dissimilar and yet they were regarded as one disease. Evidence from the Modern Pandemic suggests that both cool, damp weather and continuous exertion were probably factoring in the high level of pneumonic plague described in some of the initial plague descriptions.

Some historians have tried to explain the differences in morbidity levels between the Modern Pandemic and the Second Pandemic by arguing that the contagious pneumonic form was much more common, especially in the initial wave of the pandemic. This would help to explain the speed with which plague traveled in the initial wave of the pandemic and the extremely high morbidity rate. Although the mortality rates for various forms of untreated plague do not seem to have changed dramatically, the morbidity rate in the Modern Pandemic is much lower. Plague is also distinctive because surviving a bout of plague provides immunity for only a limited time.

The nature of this disease is important. Because plague is fundamentally a disease of animals, an enzootic, it is not dependent on a dense human population. This is in sharp contrast with some of the crowd diseases that later dominated Western Europe. For a smallpox epidemic to continue expanding and progressing, there must be 200,000 people living within a 14-day journey of one another. Unlike plague, smallpox is truly contagious in a direct person-to-person manner. Severe influenza epidemics can devastate isolated populations, but in sparsely populated areas highly contagious diseases like new flu mutations typically die out before becoming widespread. The fact that plague, unlike smallpox, measles and other highly contagious diseases, did not make it to the New World during the colonial period, despite being prevalent in England during the period of colonization, supports the argument that plague was caused by *Y. pestis*, rather than a disease spread easily from person to person.

Plague produced by *Y. pestis* is distinctly different from these highly contagious crowd diseases Europeans faced. Unlike crowd diseases, the plague is not easily spread directly from person-to-person. Rather, the plague is a vector-borne disease, primarily of burrowing animals. Its occasional spread to humans is dependent on complex interactions of independent environmental factors that allow large populations of rats, infected with *Y. pestis*, and hosting lots of fleas, to come into contact with people. The complex means by which plague is spread helps to explain why it was so difficult for Europeans of the medieval and early modern periods to understand how the plague was spread and what they could do to prevent or control it.

Plague Epidemics

Three Plague Pandemics

An understanding of the biological complexity of *Yersinia pestis'* infectious cycle can help to explain the diverse patterns of plague epidemics, especially when considered within a historical context. Communities and cultures provide different environments within which rats and fleas have varying contact with people. Also, people from disparate times and places describe their disease experiences differently.

Despite these differences, because there is so little information about many of the individual epidemics of the Second Pandemic (1347-1722), researchers often compare data from multiple outbreaks to supplement the limited information available from individual outbreaks. Patterns that have been developed by studying plague epidemics of the Modern Pandemic have been supplemented by information garnered from epidemics of the two earlier pandemics:

The Plague of Justinian (542) and the Second

Pandemic. These generalized conceptualizations of how a plague epidemic can be expected to behave can then be compared with individual epidemics of the earlier two pandemics. These comparisons provide valuable information for determining whether a specific epidemic was or was not plague and they reveal broad patterns of plague behavior. Many epidemics that occurred during the Second Pandemic are now commonly accepted to have been plague; however, an almost equal number of epidemics seem to defy the expectations developed by examining the broad patterns. While many researchers debate whether many individual epidemics were the result of plague, most now accept Shrewsbury's contention that some periods of high mortality that once were assumed to have been a plague, were not because they do not match the general pattern of plague outbreaks.

In addition to the debate about the actual cause of many individual epidemics, a few researchers also contest the definition and causation of the three pandemics, especially the Second Pandemic. Since the late 1970s, several influential researchers, including

Samuel K. Cohn, Graham Twigg, and Susan Scott, have contested this conclusion. Based largely on the evidence of the Modern Pandemic, they argue that the pattern of the Second Pandemic, especially the initial wave, does not match the expected pattern of *Y. pestis* infection.

I argue, however, that all three pandemics should be studied to develop an understanding of how plague epidemics behave, and to understand how environmental conditions and human behavior can affect the course of their development. The resultant understanding of plague patterns, however, must include three important caveats. Concrete information from the two early pandemics is severely limited, and the pattern of the Modern Pandemic has been greatly altered by changes in the human environment and human behavior and by the treatment of patients with antibiotics. Thirdly, the Modern Pandemic may still be incomplete because previous pandemics have lasted for several hundred years. Despite the lack of complete information from any one of the three pandemics, the information that does exist can be used to evaluate the observed behavior of an epidemic in light of the historically expected patterns. It is important to remember, however, that because the plague is a vector-borne disease that requires the presence of both rats and fleas, and is not a disease that is spread simply from person to person, a complicated web of factors influences the epidemic behavior of plague in humans.

The Plague of Justinian: The First Pandemic

The initial pandemic is often referred to as the Plague of Justinian. It first erupted in Constantinople in 542 and continued sporadically for several hundred years. This plague is thought to have been concentrated in the Far East, the Middle East, and the Mediterranean basin, which was the most densely populated area of Europe. It also afflicted North Africa, but little is known about its diffusion farther into Africa. The earliest Chinese records mentioning plague suggest that it did not arrive there until 610, but they strongly suggest that the Plague of Justinian produced high mortalities in China for several centuries. In Europe, the pandemic is thought to have erupted in sporadic localized epidemics for several hundred years and then it simply faded from the landscape. The extent to which this epidemic spread beyond the Mediterranean basin in the West is now debated; however, it is widely acknowledged to have been devastating within the confines of its reach, a conclusion which is supported by contemporaneous observers. Procopius of Caesarea (c.500-d.560s) wrote that "the whole human race came near to being annihilated."

Although chroniclers such as Bede (c.672-735) in England complained of epidemic diseases, historians and epidemiologists argue that neither the human nor the rat populations in Europe could have been dense enough to support a bubonic plague epidemic much beyond the densely populated coastal Mediterranean region. In the Italian region of Liguria, Paulus Diaconus (720-799) wrote about the epidemic of 566-567, which had occurred one year after many signs that had warned of the forthcoming plague: "there began to appear in the groins of men and other delicate places, swelling of the glands after the manner of a nut or date" accompanied by a fever, and many people died. Gregory of Tours (c.538-593) also mentions outbreaks of a disease he referred to as "the pestilence known as the plague of the groin." Gregory's description of the plague epidemics that ravaged the region throughout the sixth century not only includes a reference to plague buboes, it also makes note of an interval between the disease's first appearance and its major flare-up. The epidemics that

Gregory reported were in Arles and Marseille in southern France, and in Auvergne in central France at the northern boundary of Gallo-Roman France, which is at the extreme northern edge of the region that would have been susceptible to plague if human population density is a limiting factor. However, DNA from the bacillus responsible for the bubonic plague has been isolated from a sixth-century skeleton in Bavaria. This discovery of plague in a relatively sparsely populated region provides some evidence that plague had spread beyond the confines of the Mediterranean basin, and can be taken as evidence that lack of a dense human population is not a limiting factor in the spread of the plague. It also suggests that rats had spread throughout Europe by the early medieval period.

Medieval and Early Modern Plague: the Second Pandemic

The Second Pandemic entered Western Europe in 1347 via Messina at the southern end of Sicily. Evidence suggests, however, that even before it began spreading through Western Europe, North Africa, the Middle East, India, and China, the plague was epidemic in the T'ien Shan mountains of central Asia. The Nestorian community experienced heavy mortalities in 1338-39, as demonstrated by grave markers indicating mass burials. Although the point of origin of this pandemic has not yet been pinpointed, it probably was somewhere in the high steppes of central Asia. The burrowing animals of the high alpine regions are excellent hosts for maintaining a plague reservoir because their cool damp burrows provide an ideal environment for Y. pestis. The plague is still endemic among animals in the region. Traders almost certainly were central to the disease's diffusion, and because the Nestorians were traders living in the region of Lake Issyk Kul, they might be implicated in the disease's farther spread. However, there is considerable disagreement about whether this is the place of origin of the Second Pandemic.

The two areas competing for the origin point of this pandemic are Central Asia near Lake Issyk Kul (Kyrgyzstan) and Kurdistan around the Caspian Sea, but the former hypothesis has traditionally received more favor. However, because of the Medievalis Y. pestis biovar (biological variant) is currently found in Kurdistan, and because Kurdistan is between the posited African point of origin for Y. pestis and its points of greatest diffusion, this hypothesis also seems to have much to commend it. Significantly, regardless of its exact point of origin, the disease that ravaged Europe in 1348 began in a region where Y. pestis is still endemic, which thus supports the idea that Y. pestis was the causative agent.

Once the epidemic began, it spread very quickly across Western Europe. In just one-year plague spread from the southern Italian peninsula to Northern Britain and into Greece and Eastern Germany. Following its introduction in 1347-1348, plague outbreaks sporadically ravaged Europe for almost 400 years. The last major outbreak in Britain occurred in

London in 1665, which was followed by a few scattered and localized minor outbreaks elsewhere in the country. In France, the last major epidemic occurred in Marseilles in 1720; the last major epidemic did not occur in Moscow until 1771; and after a major epidemic in 1778 in Istanbul, the plague did not disappear from the Ottoman Empire until the late 1800s, by which time the third plague pandemic had begun.

The Modern Pandemic: The Third

The third or Modern Pandemic is usually described as having begun in 1894 when it appeared in Hong Kong. However, there had been sporadic localized outbreaks of plague in the previous 100 years. From Hong Kong, the plague spread across most of the rest of the world, although with much-reduced severity compared with the medieval epidemics. Even before the discovery of antibiotics, the plague reached epidemic proportions only regionally. The reasons for the change in morbidity and mortality rates are not known, but evidence suggests that genetic changes in Y. pestis is not the cause because there appears to have been very little change. This lack of change has led some historians and epidemiologists like J. F. D. Shrewsbury, Samuel K. Cohn, Graham Twigg, Susan Scott and Christopher Duncan to speculate that some other disease must be the cause of the high mortality during the medieval and early modern epidemics. Others, including William McNeill and I, argue that Y. pestis was the agent and that changes in the material and social culture are instrumental in reducing the morbidity rates in the Third Pandemic. There have been very few plague victims in Europe during the Modern Pandemic, but there have been serious outbreaks in other parts of the world. Regions as widely dispersed as China, India, Sri Lanka, Java, Madagascar, South Africa, Brazil in South America, and San Francisco in North America experienced plague outbreaks in the current pandemic. None of these epidemics has produced the widespread, heavy mortalities common in the Second Pandemic. Nonetheless, fear that the Modern Pandemic would produce mortalities similar to those of the previous pandemic has driven research on the current one.

During Modern Pandemic, many attempts have been made to control the spread of the plague by preventing the transportation of fleas. During an outbreak in Java, travelers and their luggage were inspected for fleas, but, interestingly, these inspections discovered very few fleas. San Francisco experienced a couple of epidemics in the early twentieth century that were focused on the newly arrived and impoverished Asian community. Plague first arrived in San Francisco on January 2, 1900, in Australia.

Although the ship was searched by quarantine officers and the cargo had been sanitized or fumigated, plague, presumably carried by rats with fleas, made it into San Francisco. Despite an inspection by trained agents who knew what they were looking for, the entry of plague into San Francisco casts doubt on the ability of similar measures, instituted centuries earlier, to be successful.

From San Francisco, Yersinia pestis spread slowly among native ground-dwelling squirrels, so that it is now endemic among them in Arizona, southern Oregon, and northeastern California. Nonetheless, bubonic plague only very rarely affects humans in these regions. The individual cases that do occur are almost always the result of direct contact between a person and an infected animal, the disease presumably having been transmitted by its fleas. This pattern of occasional outbreaks has been typical of the Modern Pandemic, especially in the developed world. During the Modern Pandemic, a plague has maintained a sporadic presence in India. These epidemics have occasionally been severe, killing thousands; however, morbidity rates have not entered the double digits and the epidemics have typically remained confined to a city or within a region. Ole J. Benedictow argues that one of the reasons for low morbidity rates in India is that the people have traditionally left their homes when the plague breaks out in their vicinity, rather than remaining confined in their houses near the rats and fleas that inhabited their houses as was the case in the medieval and early modern epidemics in Europe.

Although plague epidemics have occurred in Columbo, Sri Lanka, the epidemics in Columbo have been less devastating than those in India. This difference has been explained by the fact that the dominant flea species in Colombo are less effective plague vectors than those on the mainland of India. The plague has also remained a significant threat in Africa. The island of Madagascar experienced reoccurring bouts of the plague from the 1920s through the 1940s, which were extensively studied. Plague continues to be a source of concern around the world. In February of 2005, a primarily pneumonic plague epidemic that broke out in the Congo provoked fear and caused people to flee without seeking treatment, raising concerns that the epidemic may spread widely.

Despite worldwide outbreaks, only very few people in Great Britain, in several small, localized outbreaks, have been killed during the current pandemic. In 1900, there were 36 human cases in Glasgow and 16 people died, but these cases were confined to only a few houses. Rats were examined but no Y. pestis was found. In 1901, five more cases were centered in a rag store, and there were a few cases involving men who worked in two separate flour mills, where each of the men was known to have handled the bodies of dead rats. In this outbreak, Y. pestis was found in the trapped and examined rats. In 1907, another case in Glasgow involved a rag store. In the process of attempting to find the source of the disease, many rats were examined for Y. pestis; the nearest infected rat was found a mile away. These observations suggest that the plague had been endemic among the rats for at least six years without infecting any people. Also, because infected rats were not found in the vicinity of all human cases, human cases were not easily linked to rat cases.

In Suffolk, a plague outbreak began in 1906 but was not recognized as such until 1910. In 1906, eight people became sick and six died after an extremely quick onset of pneumonia. Because influenza was present, these deaths were not attributed to the plague until 1910 when additional cases occurred. This little outbreak resulted when rats and fleas carrying plague were off-loaded, presumably along with grain shipments. These twentieth-century English cases were almost exclusively pneumonic and manifested so atypically that bacterial examinations were required to prove the cause of death. The extremely limited scope of these few plague outbreaks in twentieth-century Britain has led to questions about the veracity of the plague produced high mortality rates reported in medieval Britain.

Evidence from the modern pandemic suggests that the plague is tenacious and is not dependent on humans to maintain its presence in a given locality; it can exist regionally for years without producing human victims. This evidence also suggests that relatively small differences, such as the difference in what species of flea is dominant, as in Sri Lanka/India example, can have a relatively large impact on the severity of plague outbreaks. An analysis of the different manifestations of various epidemics of the three great pandemics also provides evidence that human behavior can greatly affect the consequences of an epidemic.

British Epidemics

Plague epidemics in Britain roughly follow the pattern of the worldwide pandemic pattern with a few notable exceptions. First, as previously stated, there is little evidence that the first pandemic, the Plague of Justinian, reached Britain. In the absence of clear evidence, the debate focuses on the issue of whether Britain or more specifically England, could have provided the necessary environmental elements for the survival of plague beyond the infection of one or two people. During the period between 450-800, it is unclear whether England was sufficiently densely populated to facilitate the spread of plague, or more importantly to support a population of rats capable of harboring and dispersing plague. Nonetheless, England did experience epidemic diseases during this period even though there is a very little written description of them. Bede wrote about high mortalities and sickness, but his description of symptoms does not match very well with the plague.

Britain, like the rest of Western Europe and unlike most of the rest of the world, appears never to have had local foci of plague among native ground-dwelling animals. Britain is not home to many species of ground-dwelling mammals, although several species of wild mice, voles, ferrets, and rabbits inhabited England during the years of the Second Pandemic. During the initial years of the Modern Pandemic, *Y. pestis* was found in wild rabbits and hares as well as in domesticated cats in Britain, but these species are too susceptible to plague mortality to maintain the disease over time.

Additionally, because rabbit fleas are dependent on hormones from pregnant rabbits for their reproduction, they are extraordinarily unsuited to spreading plague to humans.

Medieval British Epidemics

The plague arrived in Britain in the spring or summer of 1348. It first arrived at Weymouth, a port in Dorset in southwestern England. In the Gray Friars chronicle, the plague is recorded as having arrived on the Feast of St. John the Baptist, June 24. The chronicle states that the "seeds of terrible pestilence" arrived on a ship from Gascony. Plague may have been present on the ship indicated by the chronicle, but for the plague to have become noticeable by June 24, it must have arrived six or seven weeks earlier. The period between when the plague first arrives in a location and breaks out in a human epidemic is the time needed for several stages to occur: the first rat infection, followed by a rat epizootic and die off; then fleas leave the dead rats and the blocked fleas begin to infect humans; and finally, the human outbreak.

The plague probably arrived about May 8 from Bordeaux in the Gascony region. Although the dates are speculative, they all fit together. The Bishops of Lincoln and York ordered preventative masses and prayer gatherings on July 25 and July 28, respectively.

The dates of these masses thus allow one month for the news to travel across the country and the decision to be made to schedule for a mass. The timing is further supported by the dates for installing priests into vacancies, assuming an average time lag of about 13 weeks between the death of a priest and the institution of his replacement. From Weymouth, the plague spread along the coast of southern England, traveling fastest along with the coastal towns and then more slowly inland. This suggests that it was carried on ships around the coast and then traveled inland along with smaller cargos on barges and in carts. The plague was first recognized in Bristol on August 15, which indicates that it had arrived there several weeks earlier.

The plague spread throughout southern England in 1348, through central England in 1349, and into northern England and Scotland in 1350. The date plague was first recognized in London is uncertain because two dates, Michaelmas, (September 29) and All Saints Day (November 1), have been associated with its introduction.

In Britain, a great deal of the statistical

information on the initial epidemics is based on clerical records, principally on the number of priests who were appointed to positions and instituted to benefices, appointed presumably to replace priests who had died of the plague. There has been a debate about how representative these figures are of the mortality rates in the general population. This argument has been fueled not only by the high mortality level of priests, given that at least 45 percent of parish priests appear to have died in the epidemic, but by how relatively low the mortality figure for Bishops appears in contrast, only 18 percent.

The idea that general mortality levels in Britain could have approached 45 percent has been resisted.[33] It has been argued that priests because they were on average older than the population at large and had duties to perform for the ill and dying, putting them at greater risk, had unrepresentatively high mortality figures. Additionally, Shrewsbury argues that the priests' mortality figures appear higher than they were because many priests were appointed to benefices that became empty due to flight rather than mortality. A letter by the Bishop of Bath and Wells addressed to his priests says, that "since no priests can be found who are willing, whether out of zeal or devotion to exchange for a stipend, to take pastoral care ... to visit the sick and administer to them the sacraments of the church," when necessary, anyone, even a woman, can hear the last confession. This suggests both that the crisis was severe and that priests were fleeing their duty. Scholars have resisted accepting priests' high replacement rates as indicative of their mortality rates, let alone as representative of the overall mortality rates, in part because they are so high. Their resistance to accepting these figures has been supported by the great variation in the mortality rates of various religious communities.

In contrast, the mortality rates reported for priests were probably lower than that for the general population because priests would have been better housed and fed than most other people, and they would have received care from the community as a whole rather than just from their immediate, and probably also ill, family members. Additionally, the record of complaints that priests were not as attentive to the needs of their parishioners as they should have been suggesting to some scholars that priests' risks of contracting plague were no higher than the population at large. From this perspective, it has been argued that mortality figures inferred for the general population from the number of priests instituted may be unrepresentative low. Benedictow examines this issue in great detail and presents a very convincing argument that basing general mortality rates on the institutions of priests into empty benefices underrepresents the actual mortality.

He argues that curates and lay workers, whose replacements were not recorded in the Bishop's lists of newly instituted priests, took on the parish tasks that provided the highest risk of exposure to plague. He also points out that during the height of the chaos created by the plague, many benefices with low stipends probably remained unfilled until after the epidemic. The number of priests instituted remained high through 1350 and then returned to normal. Benedictow also notes that in the absence of antibiotics, the one factor that can improve one's chances of surviving plague is good palliative care. He argues that parsons, who were desperately needed by their parishioners, were more likely than members of the general public to have received good care. Because of the way plague is spread, it was centered in individual households, with the result that families struck by plague often had no good person available to provide care for the ill.

In the initial wave of the Second Pandemic, enough people became ill that the normal means of care-giving by members of the immediate household became interrupted. High morbidity, coupled with fear, exacerbated the death toll. Taking all of these somewhat speculative and contradictory factors into consideration does not allow for an exact estimate of the level of mortality during the initial outbreak, but it does suggest an idea of probable mortality levels. John Hatcher concludes that "A national death-rate of below 25 percent or above 45 percent would appear most unlikely." However, researchers have more recently begun to accept the priests' reported mortality level as a minimum mortality level rather than a maximum; Benedictow has even concluded that the mortality rate in England may have exceeded 60 percent.

Epidemics in Summary

Although the three pandemics vary greatly, they also share many similarities. The first two great pandemics encompassed a large portion of the Old World. In both, mortality rates were so high that at least some people feared that the society they knew would not survive. Although the Modern Pandemic is more widely distributed, mortality levels have never reached a point that has provoked panic except on a very local level. Of course, when cases of plague were identified in Hong Kong, the fear provoked by historical records of the Second Pandemic drove the decision to respond immediately to the outbreak proactively and decisively. The first two pandemics extended across four centuries, and the duration of the Modern Pandemic is unknown, but it has already lasted more than a century. Thus, plague contrasts sharply with other epidemics such as that of the highly contagious Spanish flu (1918-20) that quickly spread around the world and then disappeared.

The initial wave of the Second Pandemic exhibited a distinctively rapid pattern of movement coupled with high mortality that can be traced on a map. Subsequent outbreaks of the more than 300-year pandemic seemed to pop up sporadically, and they only occasionally produced very high mortality rates. Despite these differences in the way plague was distributed across Europe and variations in how it presented, contemporaneous observers perceived the outbreaks as one disease. The stability of the pathogen itself does not imply a consistent impact either on a rat or human populations. Plague episodes were modulated by many different factors.

London

London was an atypical city, by far the largest, most diverse, and busiest city in Britain, however, London and its environs can serve as a microcosm of the English plague experience. Many of the individual changes observed in London can be found throughout the rest of England. According to records of the late sixteenth century and the seventeenth century, the plague had a nearly constant presence in London, while in other parts of England plague outbreaks were more sporadic. Although plague was virtually always present in London, in most years the disease was of limited significance, both in terms of the numbers of people afflicted and in terms of the affected parishes. London's experience with the plague was not unique but it was more condensed and intense than in the rest of England and thus it is a good city in which to examine the plague and its cultural impact. Extensive records exist, especially from the last hundred years of the epidemic, and Londoners were forced by the virtually continual presence of plague to develop methods of coping with the disease.

To consider possible mechanisms for the distribution of plague across London, the social and physical geography of the city should be examined. Especially important are city features that may have harbored or facilitated the dispersal of rats or brought rats into close contact with people. Plague is often described as a disease of place. Once an environment is infected with *Y. pestis*, plague becomes endemic. Today plague is present in vast regions across the globe although few people are infected. The infection of place can extend from an individual house or building to the macrocosm of a village or city, for example in India when the plague breaks out, villagers leave their homes to camp in the open air and thus avoid the worst of the outbreaks. During the Second Pandemic, the plague's association with specific places is less clear than it has been in India but by the fifteenth century, Londoners knew the best way to avoid the plague was to leave London. It is important to examine the geography of historic London because of changes during the intervening years, especially damage caused by the Great Fire in 1666 and extensive development during the intervening 450 years.

London, England's principal port city, developed along the River Thames at the site of the London Bridge, first built in Roman times and the only bridge across the Thames for many centuries. London is about 35 miles inland from the ocean, and because the Thames is a large, navigable, tidal river, it provided a safe port for shipping with easy access to an urban population. The majority of London docks were located just slightly east and downriver of London Bridge and central London; however, throughout London's formative years it could be raised to allow ships to reach upriver docks and Westminster (see Figure 1).

Figure 1. Map of London showing rivers, wall, gates, parish boundaries, Westminster, and Southwark. Map of parish divisions adapted from Justin Champion.

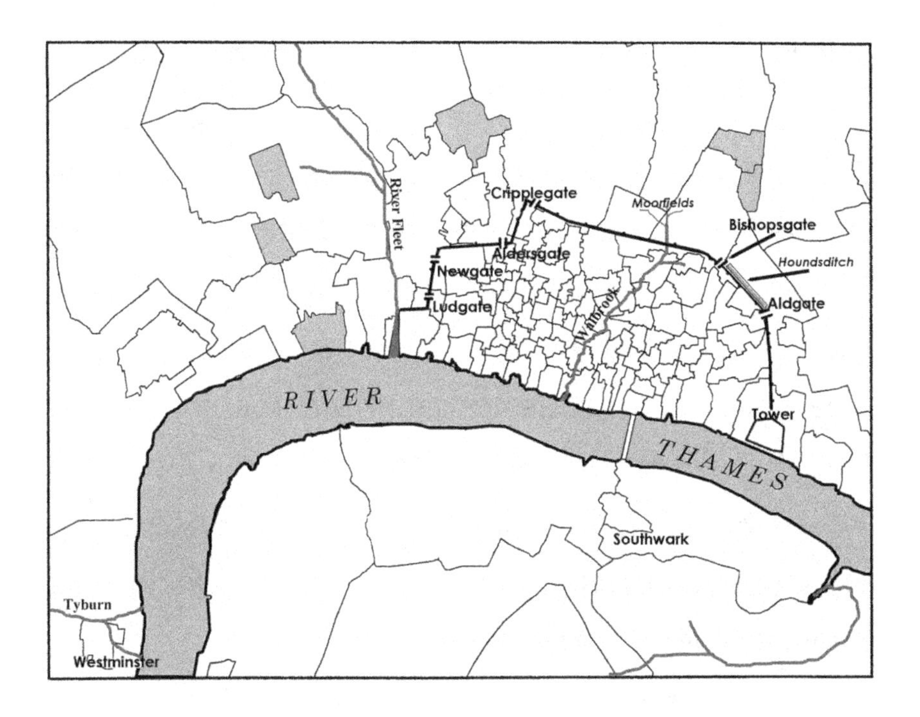

That London was a significant port city is particularly relevant in a study of plague because as a port city, it was the point of entry for goods from the rest of the world, material that potentially harbored either or both rats and fleas. Thus, it is necessary to consider London's history and its geography to reconstruct the environment where the plague thrived. Medieval London was also traversed by the Fleet and the Walbrook rivers as well as by their tributaries. All of these rivers served as waste removal systems and thus were both a source of food for rats and a barrier separating communities of rats from one another. The Fleet River entered the Thames immediately outside the western Roman wall and east of St. Brides, which was the site of Henry VII's palace of Bridewell. In the medieval period and even into the early modern period, the Fleet was a real river that experienced tidal flow; however, when John Rocque produced his London map in 1746, he labeled it as the Fleet Ditch. Centuries earlier, there had been periodic complaints that the Fleet stank or that its flow was blocked by garbage.

The path of the Walbrook is now impossible to

see because it flows in conduits 32 feet below street level; however, in the early fifteenth century, the Walbrook was an open river used as a free-flowing sewer. After years of only sporadically enforcing rules that forbade latrines that dumped into the Walbrook, in 1383 these latrines were legalized with stipulations that forbade rubbish that blocked the river, and the latrine owners were charged a usage fee to provide for cleaning out the river bed. When the Walbrook was free-flowing, it emptied into the Thames at a place known as the Dowgate. In 1462-3, the Common Council ordered landholders along the Walbrook to cover the stream and close up the latrines that emptied into it; the Walbrook does not appear on sixteenth-century maps. Nonetheless, the Walbrook served to drain an area known as the Moorsfield, a wet and boggy region just north of the city walls between Cripplegate and Bishopsgate, an area which was often flooded, and was suitable for boating at times, and was the site of ice skating in winter. So presumably, water must have continued to flow in the Walbrook riverbed even after it was covered by construction. In 1511, the mayor ordered dikes to be built so that Moorsfield could be more easily crossed, and the area was eventually filled, raising its elevation about 12 feet; nonetheless, sections of Moorsfield remained undeveloped open fields into the eighteenth century.

London was established in 50 AD and the wall that gave medieval London its distinctive shape was in place by 200 AD. It was a defensive wall, surrounded on the exterior by a moat or wet ditches. On the western side of London, the wall was bordered by the Fleet River, but elsewhere the wall was bordered by moats that had been dug as an additional barrier against attack. Within the city, the wall was bordered by open space. As London grew, the ditches were often used as dumping grounds for trash, debris, and human wastes. One section of the ditch between Bishop's Gate and Aldgate on the east was referred to as Houndsditch because of the stink generated by the garbage and dead dogs tossed into it. These ditches, where garbage and food wastes collected, might have increased the rat density outside the walls, and thus they might have played a role in changing the plague locus from within London walls to the suburbs without the walls. Also, the open space within the walls would have served to segregate the urban rats from those outside the walls in the suburbs. Such segregation undoubtedly would help to explain why, in the later London epidemics, the plague was focused in the suburbs outside city walls and ditches.

Beyond the wall, the greater London area included the Tyburn River, the main portion of which entered the Thames south of Westminster, while a second branch flowed around Westminster to the north, making Westminster an island. However, by the late sixteenth century, the north channel of the Tyburn appears to have been built over. The greater medieval London area also included the city of Southwark, on the south side of the Thames just across the London Bridge. The land west of Southwark was marshy and uninhabitable until it was filled in and built up and as the Thames became channelized.

Although London was never a particularly hilly town, its topography has been leveled out by human activity during the hundreds of years of its existence. Hills were reduced, marshy regions were drained and filled, and Londoners built up the river banks and expanded the city into what had been part of the Thames channel. In general, London's topography slopes toward the banks of the Thames, but the landscape is complicated by slight dips in elevation toward the banks of the Walbrook and the Fleet. As these rivers were built over, passages large enough to allow rats to move along the underground chambers undoubtedly remained. Black rats, *Rattus rattus*, sometimes referred to as ship rats were the only rats in Britain until the 1700s, when Brown rats, *Rattus norvegicus,* were introduced. Black rats are climbers that today are more likely to be found in rafters rather than along underground rivers or garbage sites; however, little is known about their behavior before they had to share their environment with the bigger, fiercer, ground-dwelling Brown rats.

London's rats inhabited a city that included large blocks of interconnected houses built from the debris of older buildings a type of construction that would have assured easy access for rats throughout an entire block of houses. These housing blocks were of irregular shapes and sizes, the blocks were penetrated by narrow alleyways and separated by open public streets, which would have served to separate rat populations of one block from another. Although the physical geography of London did not provide humans with the same difficulty that it would have provided rats, the complicated social geography of London certainly affected the way that plague was combated throughout the city. When the plague first arrived in England, in 1348 London already had developed distinct patterns of governance that affected its struggle to combat the plague. London was granted a charter in the early twelfth century that gave it responsibility for its governance as well as freedom from the direct control of the monarch, and by 1348, the geopolitical landscape of London was a complex network of interconnecting and overlapping jurisdictions. These produced complexities that can be seen in the multiple levels of reporting and oversight necessary to record plague deaths.

Within its walls, London, which comprised just one square mile of land, was composed of 97 parishes. Outside the walls were another 16 parishes that were within the City Liberties and thus under the direct control of the corporation of London. London also included areas within City Liberties that became available for city construction when Henry VIII broke up the religious houses. In addition to being divided into parishes, the areas within the jurisdiction of the London Corporation were divided into larger political regions known as wards each of which were headed by an alderman.

Parishes oversaw many social services and charity functions for the indigent, including care of plague victims and monitoring plague deaths, while wards managed other governmental functions, including enforcement of plague orders. In addition to the areas under the jurisdiction of the City corporation, metropolitan London included Westminster and Southwark. More importantly in terms of urban management, it also included a growing number of outlying parishes, with growing populations, which were self-managed but overseen by the system of justices of the peace that managed the rest of the country. During times of crisis, these outlying regions of greater London received special directives from the Privy Council.

During the Second Pandemic, metropolitan

London was a region where plague control was directed by two sets of plague orders, Royal and City; further, these orders were enforced and supported by a diverse group of entities. Fortunately for metropolitan London, the Royal plague orders and those issued by the London corporation or the mayor were virtually identical; however, because of the complications of the multiple political subdivisions, it is unclear how the plague orders were applied and it is unlikely that they were applied or enforced evenly within jurisdictions, let alone across jurisdictional borders. Within the City of London, Aldermen, were responsible for the enforcement of orders. The aldermen were by mandate substantial men who undoubtedly lived on wealthier streets, and who were thus likely to enforce plague control measures more strictly in their neighborhoods than elsewhere. Properties within the London Liberties were subject to stringent rules of housing construction that were not applied outside City Liberties, although laws enacted during Queen Elizabeth's reign indicate that attempts were made to control construction and subdivision of multi-individual housing in the suburbs.

As London grew and developed, old buildings were demolished and the rubble was used in new construction. Early medieval buildings were made of rubble from the Roman city, and this rubble from old buildings was reused as buildings were continually torn down and recreated. This continual rebuilding and reconstruction of buildings and sites have given the development of London an organic or almost haphazard organization and created groups of interconnected buildings porous enough for rats to move through them.

London buildings were commonly built side by side, abutting or sharing sidewalls and facing onto the street; however, buildings were frequently pierced by a narrow open-air passage that led into a backspace, or courtyard. Kitchens were often located off the courtyard rather than indoors to reduce the risk of fire. Some buildings had cesspits for disposal of garbage and human wastes and these were often placed behind the building or in their undercrofts. Many buildings were subdivided into multiple tenancies so that some tenants' only access to air was from the courtyard. The crowded condition of some of these tenancies was considered to be a risk factor for the plague as indicated by comments and complaints, and by orders that forbade the division of houses into multiple tenancies. Despite the rules, as the population of London continued to grow, houses continued to be subdivided, as evidenced by the need to reprint and reemphasize these Elizabethan laws in 1636. Among the many causes of a plague that Kellwaye mentioned are "shutting a great companie of people into a close, narrow or straight roome, as most commonly we see in shippes, cõmon Gayles and in narrow and close lanes and streetes, where many people doe dwell together."

Many rules concerning the construction of buildings within London, like the buildings themselves, developed over time. The first set of documented rules for buildings was instigated during the period of the first Mayor of London, Henry Fitzailwyn (1189-1211), and include regulations requiring the walls partitioning property lines to be made of stone. Additionally, rules were governing the thickness of shared walls. In the late twelfth century these walls were mandated to be composed of stone three feet thick, although this later was reduced to two feet. Also, rules drafted around 1200 were intended to reduce the risk of fire within London by prohibiting thatched roofs. Placement and construction methods for cesspits were similarly regulated; they were not allowed closer than two feet from a building's wall, and after about 1200, all new cesspits were required to be lined with stone whereas earlier pits had often been linked with woven sticks and clay. In the fourteenth and fifteenth centuries, probably, the primary use of stone in the construction of ordinary houses was for cesspits and by the fifteenth century, many houses had pipes or funnels leading from the privy to the cesspit.

Despite rules regulating their design and placement, complaints and suits were common, both because the rules were not always obeyed and because the rules and the construction techniques available were not sufficient to prevent occasional problems. Although the foundation walls at property boundaries were mandated to be built of stone, the bulk of buildings were made of timber, lath, and plaster, and thus were often in need of repair. By 1610-12 when Ralph Treswell performed a survey of

London houses, no houses had four stone walls and only a few had any. As London's population grew, the rules that were primarily to address property rights, fire, and general construction safety, were augmented with new ordinances created to restrict urban development and thereby to limit population growth. The increasing numbers of day laborers and transients were regarded as a public threat, because they produced the potential for social instability and because they posed a public health threat. The final item in the plague orders first issued during the 1578 epidemic, "the preservation of her subjects who by very disorder, and for lacke of Direction Do in many partes wilfully procure the increase of this general contagion," makes this argument rather subtly.

London plague orders issued in 1608 make the connection between transients and plague more bluntly: "for as much as nothing is more complained on then the multitude of Roagues and Wandering Beggers, that swarme in every place about the Citie being a great cause of the spreading of the infection." The growing number of laborers required an increasing number of low rent accommodations, which were created by dividing houses into multifamily dwellings. These multiple tenancies were seen as a source of urban danger not only because they provided a respite for rogues and vagabonds who disrupted civic functions and tested the limits of parish social services, but also because they were linked to outbreaks of plague. The tenements are described as constantly expanding so that a "infinite number being pestered together breeding and nourishing Infection so that the same tendeth to the great imminent danger of the government and safety of this Citie."

It is unclear to what extent construction rules were followed. Thatched roofs were easily visible and were replaced as ordered; however, because it is much harder to see foundation walls or to investigate the construction or placement of cesspits within private courtyards, householders may have gotten away with not obeying the laws. Disobedience of some rules, essentially sumptuary rules for housing, can be traced by the records of fines paid and assessed for their infractions. Especially egregious problems, such as a cesspit that had leaked into a neighbor's house, occasionally appeared in *Assize de Nuisance* proceedings. However, many buildings continued to be subdivided, and evidence of this exists in the plague orders printed in 1636, which include a reprint of Elizabethan material as well as contemporaneous court cases against tenement landlords. London's population growth of laborers necessitated an increase in tenements, "it was a market response to a social need that the authorities refused to deal with through public policies." Despite Queen Elizabeth's laws and those enacted,

> *since his Majesties most happy Reigne, and ... Orders and Decrees taken in this honourable Court for the restraining and reforming of the multitude of new erected and divided Tenements, and taking in of Inmates, yet nevertheless the same doe so daily increase and multiply in every place in and about this City of London and the Suburbs thereof.*

Although the presence of large numbers of poor people made the wealthier people nervous, wealthy people owned properties as investments. Rents could provide significant income, so wealthy landlords broke the law and subdivided houses despite society's concerns about the dangers of overcrowding. Further, there was no simple solution to housing London's poor, therefore the law did not require landlords to tear down houses full of people, although city officials saw these tenements as dangerous. Instead, regulations enjoined landlords from accepting new tenants, and required them to pay taxes "towards the finding and maintaining of the poore of the Parish in which such Buildings are."

In 1529, Henry VIII gave up his palace of Bridewell, located immediately west of the Fleet, due to the stink coming from the river and fear of the plague. Bridewell Palace became a workhouse and thus housed many indigent people in one building in an area that had been threatened by the plague even before it became inhabited by the destitute. Some documented reports indicate that despite the regulations, some houses had a large number of occupants. For example, in 1603, the Recorder of London noted 800 plague cases in one building that housed 8,000 inhabitants. Although the 10 percent mortality rate for plague in this one building was low compared to the general London mortality rate that approached 25 percent, the sheer number of plague victims within one house was seen as evidence of the dangers of overcrowding. Over time, some rules were changed to align with actual practice, like the rule prescribing three-foot-thick foundation walls, which was officially reduced to the two feet width that had become the norm. Rules concerning multiple tenancies were amended so that landlords were required to pay taxes to the parish in which the tenement existed, regardless of where the owner lived, to offset the costs of poor tenants to the parish. Other rules may simply have been ignored, except under egregious conditions. Plague orders that were repeatedly reissued during times of plague always emphasized the need for cleanliness, suggesting that urban cleaning was considered less important in the periods when there were no serious plague outbreaks.

London of the medieval and early modern period had become a complex web of semi-independent but linked environments, creating conditions that facilitated the spread of the plague. Rats and humans shared an environment that fostered the development of separated, relatively stable rat colonies. The environment constructed by humans served both to nurture and separate these rat colonies and yet as people moved between these colonies and changed the distribution of food and goods, human behavior served to spread the disease among rat colonies. Further, the divisions that defined the complex social geography of London and which played a dominant role in combating the spread of plague did not necessarily correlate well with the actual distribution of infected houses and plague outbreaks.

London Plague Epidemics

Records from the initial plague outbreak in London are so confused that it is not possible with any certainty to determine the date when the plague was first recognized in the city, however, throughout the pandemic, records for London become increasingly detailed. Also, London's proximity to Westminster and Whitehall, the permanent seat of royal and political power, assured London a special place in responding to national problems. Plague Orders issued in 1636, for example, stated that the current plague epidemic was putting the King and his heirs at risk.

Royal orders addressing matters of behavior during periods of epidemic plague closely mirrored those issued by London, with a few minor exceptions. In the provinces, which were governed by Royal orders, restrictions on travel and on leaving the confines of a plague-infested house were more lenient than those mandated in London. In regions outside London, where houses were widely spaced houses, the allowance was made for people living in infested houses to move about, especially in the evening. This freedom allowed people to get necessary supplies in small villages, where there were not enough people to run errands for those confined in quarantined houses and to care for their crops and animals. One additional difference between the London orders and the Royal orders was that those issued by the Crown mandated that all people known to have been exposed to plague carry white staves, while the London orders mandated red staves.

When the plague entered London in 1348, the city had a population of approximately 50,000, which was almost certainly twice the population of the second-largest city in England. According to Slack, by 1550, London's population was approximately 85,000, having almost tripled during a period when the population of the country as a whole barely regained its pre-plague population. As London grew, an increasing percentage of the population was living outside the walls and beyond City Liberties. This growth increased the pressure to subdivide large houses into multiple tenancies. The London metropolitan population grew to approximately 141,000 in 1600 and it had again tripled to at least

400,000 to 459,000 by the time of the Great Plague in 1665. These figures are estimates, however, and those made by other researchers vary considerably.

Paul Slack notes that it has "become a truism that there was scarcely a year in the sixteenth and the early seventeenth century" when the plague was absent from London. He is skeptical of the mortality figures provided by the London bills of mortality in years when very few deaths were attributed to the plague. Because people wanted to avoid having their houses quarantined if they possibly could, it seems likely, especially in years with few recorded plague deaths, that the number of plague deaths listed would have been under rather than over-reported. A few scattered human deaths are exactly the pattern of deaths to be expected as the result of occasional human contact with the fleas of infected rats in an environment of many independent but interconnecting rat colonies. The tenacity of plague epidemics in London over three centuries suggests that plague was enzootic in London's rats during these 300 years, although historians including Paul Slack and J. D. F. Shrewsbury have assumed that plague was reintroduced before major outbreaks. That there were years with very few plague deaths reported strongly supports the argument that the epidemics were caused by a disease that was not easily communicable from person to person.

The dynamics of endemic plague are such that infected fleas would regularly give plague to only a small number of people, and only in years of epizootic plague in rats did the disease become epidemic among humans. These epizootic outbreaks of plague could be triggered by any number of environmental factors, such as the introduction of more *X. cheopis* fleas or the introduction of a new strain of *Y. pestis*. Anything that affected the distribution of *Y. pestis*, and its hosts or the contact they have with humans would affect the distribution of human plague outbreaks. Because black rats are unlikely to travel beyond the confines of the building in which they are born, the housing pattern of London with units of linked buildings separated by streets would have allowed for a large number of independent rat populations that would only occasionally come into contact with one another.

In Tudor and Stuart London, mortality figures

were collected at the parish level by way of two different channels. Burials were recorded in parish registers at the time of burial, and these records often mention the cause of death or provide information about the victim's occupation or status. During periods of high mortality, the records are substantially less complete, either because the information was unavailable or because the parishes did not have the personnel to properly record information on the large numbers of burials they were forced to perform. During these periods, there are anecdotal reports of bodies being unceremoniously left at a church for burial with no record of who they were. In the register of St. Botolph Bishopsgate, records from 1563 include notations such as "two corpses" or "three corpses." In 1625, the clerk at St. Saviour Southwark "sometimes found twenty or thirty corpses left at the place of burial."

The second set of official records formed the material printed in the London bills of mortality. These were compiled from the data collected by parish workers, who were referred to as searchers or viewers. By mandate, they were women of "the best sort as can be got in this kind" to fulfill the position. Despite regulations requiring that parishes hire only the best available women, Paul Slack notes that these women have often been referred to as "ignorant and careless searchers" and that their reports have been challenged. Because the word of poor elderly women was not highly respected, the work of these women has been open to substantial amounts of criticism. Slack himself places an emphasis on the searchers' fallibility and John Graunt (1620-1674) complained of their incompetence. The 1636 and 1665 plague orders specify that supervising physicians should assure themselves that the searchers are doing their jobs competently. Notwithstanding the complaints, this position, created as a temporary one to provide data for the occasional bills of mortality printed during plague epidemics, became a permanent position when the bills of mortality began to be printed regularly. The people who held the position of searchers, despite being 'ancient matrons' rather than trained medical practitioners, were the primary sources of statistical data on the causes of death until 1836.

Searchers went to the houses of all reported

deaths to investigate the cause of death, and they reported their findings to constables who in turn reported the information to parish clerks. From the individual parish clerks, the data were sent to the Clerk of the Parish clerks where it was compiled to form the Bills of Mortality.

Positions as searchers were given to needy but able-bodied parish women, often widows, and in effect were work demanded in return for the charity provided. Plague orders stated that a woman who refused the role of searcher could lose her pension. There is little doubt that these women were not in a position to withstand the pressure applied to classify deaths in the households of the wealthy and powerful as being due to anything other than the plague. However, these women also were required, under penalty of punishment, to report the truth. Item 4 in the plague orders first published in 1578 states:

> *And in case the said views either through favor or corruption, shall give wrong certificate, or shall refuse to serve being thereto appointed, then to cause them to be*

punished by imprisonm't, in such sort as may serve for a terror to others.

Both the national plague orders and those issued by the Lord Mayor and Alderman addressed concerns that the searchers were inept by requiring that physicians who were ward officials appointed to deal with the plague, and the surgeons who supervised the searchers, assure themselves that the searchers working with them were qualified for their task.

If mortality reports were indeed inaccurate or altered, there is no reason to limit the errors or fabrications as the sole responsibility of the searchers. Householders who had no desire to have their houses shut up had very good reasons to pressure both the searchers and parish authorities not to report deaths as the result of the plague. However, reports could have been altered because of pressure applied to one of the many other points in the data collecting process before being printed in the bills. Parish officials with more prestige than searchers had more to gain by under reporting the number of plague deaths in their parish. Indeed, Pepys notes that the clerk of his parish admitted to him that nine people in the parish had died of plague but that he had reported only six plague deaths to the Clerk of clerks. This may suggest that parish clerks in the wealthy parishes were more adroit at misreporting the causes of death.

As significant as the purposeful under the registration of plague deaths may have been, an equally significant reason for the under-count of plague deaths in the bills of mortality is that plague was sometimes difficult to identify. Steven Bradwell's 1625 plague treatise, *A Watch-man for the Pest*, describes the symptoms of plague in terms that would not have made its recognition easy. He says, "As soon the Heart is stricken with putrid vapor, the spirits grow distempered and inflamed. And this distemperature is a feavor (not Proper, but Symptomatic or Accidental) and the Feavor is not of one kinde in every one; but diverse." He also describes the theoretically more obvious outward signs of the plague:

a secret sinking of the Spirits and

> *Powers of Nature, with a painful wearinesse of the bones, and all without any manifest cause. Then follows great trouble and oppression of the heart, that the partie unquietly rowles up and downe for rest from one place to another sighing often.*

According to Bradwell, these symptoms were followed by attempting to vomit or vomiting "filthy stuff of various colours" and then by head pain and faintness. It was only after these symptoms of plague were noticed that the most obvious tokens of the plague would become apparent, and he further notes that these surest signs are not apparent in all plague cases. It is clear from this description of plague symptoms that even the most educated people could have trouble differentiating plague from other maladies. Nonetheless, people did distinguish plague from other illnesses, although it also is clear, as John Graunt points out, "that in the Years of Plague, a quarter more dies of that Disease than are set down." The use of crisis mortality levels, in which the total number of deaths reported during plague epidemic years is compared with the total number of deaths reported during normal years, allows historians to ignore the problem of under-reporting of plague deaths. Even in the twentieth-century, plague cases have been misdiagnosed, or as in the small plague outbreak in twentieth-century England, only diagnosed after a bacterial investigation was performed. Certainly, 400 years ago, it was much more difficult.

Evidence suggests that plague was endemic in London for several hundred years although it only sporadically erupted in serious outbreaks; occasional years of substantial mortalities were interspersed between years with only limited mortalities due to plague. It is clear from the data available from the final 100 years of plague in London that the heaviest mortalities shifted from one parish to another with each epidemic outbreak. Unfortunately, information available from earlier outbreaks does not provide enough detail to determine whether the early outbreaks exhibited similar patterns of movement.

Both biological and environmental factors are responsible for changes in the distribution of parishes most severely hit by the plague. Some of the changes in plague distribution likely were due to fluctuations in resistance to plague within different rat populations; however, innumerable environmental changes could also have contributed to these fluctuations, including changes in the built environment and residence patterns. The human population of London grew dramatically during the 350 years of plague and in the process, the human population density of individual parishes changed.

Also, the relative density of the parishes changed; over time an increasingly large proportion of Greater London's population resided in the larger extramural parishes, which lost much of their rural character. Most importantly, the distribution of the later outbreaks of plague is unambiguously skewed toward poorer parishes. Although changes in population may have affected where plague outbreaks appear to have been centered, changes in behavior and living arrangements that were made possible by increasing wealth in London may also have influenced the plague distribution. Much of the existing information on mortality rates before 1538, comes only from wills, and since the very poor seldom wrote wills, it is difficult to determine and compare relative mortality rates between rich and poor. Richard Britnell notes that the number of wills enrolled between Michaelmas 1348 and 1349 was 111, while the average number in the previous 20 years had been fewer than three a year. Although plague was more common among the poor, a fact that was recognized during the entire epidemic period, it struck down wealthy people as well. Simon Kellwaye, writing in 1592, specifically noted that the disease is such that in some "yong and olde, rich and poore, noble and ignoble" the disease cannot be overcome no matter what means are tried. Thus, plague outbreaks continued to frighten the wealthy, many of whom fled infected areas when plague outbreaks occurred.

During the final 102 years of the plague's residence in London, there were seven notable, severe epidemics, in 1563, 1578, 1593, 1603, 1625, 1636, and 1665. In absolute numbers, the final epidemic was by far the largest, but in terms of the percentage of the population killed, the epidemics of 1563 and 1603 were the most severe. The relative severity of these two outbreaks is difficult to determine, however, not only because so little is known about the base population and mortality of London during this period, but also because bills of mortality only began to be regularly kept in 1600. Thus, the mortalities of neither of these epidemics can be compared with averaged previous mortalities. After 1603, the records are complete enough that it is possible to compare deaths recorded during plague years with those in more normal years. Brief examinations of existing data on these seven plague episodes, even with the deficiencies in data, can provide an overview of the patterns and suggest the kinds of factors that influenced the distribution and severity of plague outbreaks. These findings provide an important clue as to how the plague spread and how the human environment and behavior influenced its spread.

Once an outbreak began, typically it only slowly became distributed throughout the city. Although few parishes completely escaped any of the epidemics, outbreaks took months to spread throughout the city. Often the disease popped up in one house but not adjacent houses, and within London, it cannot be shown to have been distributed from market places. Despite high mortalities, the patterns of distribution strongly suggest that they were produced by a disease that was not highly communicable from person to person.

Before the mid-sixteenth century, much of the information about epidemics is gleaned from records of individual parishes, anecdotal reports, and records of proven wills. Information obtained from proven wills shows that there were several periods of high mortality in the first half of the sixteenth century, and all but one of these peaks occurred at times when there was contemporaneous concern about the plague. The juxtaposition of increases in proven wills and public discussion of disease is suggestive, but it does not provide a causal connection between the upswings in proven wills and plague or any other specific disease. The first half of the sixteenth century did see epidemic outbreaks, many of which almost certainly was the plague, along with the English sweat and other diseases such as typhus. It becomes possible to compare the record of registered wills with burial records for the period after 1538. The data from these two sources correlate fairly well, although there is an increase in burials recorded in 1548 that has no corresponding peak in the record of wills proved.

In the period before improved data collection, statistical data are overshadowed by stories of plague and how the fear of plague affected behavior. Plague drove Henry VIII (1509-1547) out of London many times as well as causing him to change his travel plans. In 1526, the Venetian ambassador died in London, possibly of plague, and Henry VIII spent the summer away from the city; during an epidemic in 1531, he stayed at Hampton Court. By this period, Shrewsbury claims, contemporaneous observers recognized that plague was more devastating among the poor; however, he is equally emphatic that people were not precise in their use of the term plague and he points out cases, where the plague was used to refer to an epidemic of what he believes, was some other disease. After epidemics in the late 1540s, London appears to have experienced relatively few plague mortalities until 1563. Although more complete records began to be kept during the later sixteenth century, the bills of mortality that record the major epidemics of 1563 and 1593 exist only as later copies.

In Conclusion

London's history, growth, and complex political development affected and were in turn influenced by patterns of plague proliferation. The changes London experienced exemplify the English national plague experience except that in London, unlike other English towns, the plague was virtually always present. The continuity of London's plague data allows an investigation of how patterns of plague distribution varied over time.

Seven Epidemics in London's Last 100 years of Plague

Epidemic Years	Plague Deaths	Deaths	of Deaths reported caused by the plague	Population Estimates	Index of relative mortality
City and Liberties					
1563	17,404	20,372	85	85,000	24.0
1563*	17404	23,660		80,000-	
1578	3,568	7,830	46	101,000	7.8
1593	10,675	17,893	60	125,000	14.3
City, Liberties and Nine Out Parishes					
1603	25,045	31,861	79	141,000	22.6
1625	26,350	41,312	64	206,000	20.1
City, Liberties, Nine Out Parishes, and Seven Distant Parishes					
1636	10,400	23,359	45	313,000	7.5
1665	55,797	80,696	69	400,000 - 459,000	17.6
City, Liberties, 16 Parishes Without, 12 Outer Parishes and Westminster					
1665*	68,598	97,306		500,000	

Table 1. London Mortality in Seven Epidemics

— 1563 —

The outbreak of plague in London in 1563 is the city's earliest relatively well documented plague outbreak. It was also one of the most severe, but because bills of mortality were not regularly kept before 1560 there is no baseline mortality to compare the 1563 epidemic mortality levels against; thus, it is impossible to state categorically that it was the worst of the epidemics (see Table 1). The first recorded plague death of the year was in the parish of St. Michael Cornhill (74) on March 27, yet the first death in St. Andrew Holborn (98), was not recorded until July 23, and the parishes of All Hallows Honey Lane (5) and St. Pancras Soper Lane (88) reported relatively little increase in mortalities. Despite the differences in the dates when parishes registered their first plague deaths and differences in plague intensity, deaths citywide gradually escalated until they reached a peak of more than 1,800 deaths in the first week of October. Deaths then tapered off, more rapidly than they had begun, in the late fall and winter. This is an example of the classic plague epidemic pattern in which plague deaths increase slowly at first and then more rapidly until the height of the flea season, and then rapidly diminish as cold weather reduces the flea population.

According to Paul Slack, the 1563 plague outbreak is notable because it was the last London epidemic focused in the center parishes within the city walls, rather than in the parishes beyond the walls. During this outbreak, unlike subsequent epidemics, plague seems to have struck wealthier parishes with the same severity as it struck poor parishes. Moreover, according to Slack's crisis mortality ratios, it appears that the poor parishes, which were located primarily at the periphery of London, along the Thames in the south, and to the northeast just within and beyond the wall, had the lowest mortalities of all the parishes. Slack also points out that after 1563, the poorer outlying parishes grew more rapidly than the better established, wealthier parishes within the walls; therefore, in the later epidemics there were many more potential plagues victims in poor parishes. He attributes the increasingly high morality in the poor parishes to an increase in population density that allowed the plague to spread quickly and easily. Slack also points out that even in the earliest epidemics the poor may have suffered higher rates of mortality than the wealthy but he emphasizes that after 1563 this distinction in mortality rate is visible not just in individual households but at a parish-wide level.

Although there were no unified national plague regulations before 1563, towns across England began enacting regulations to prevent people from leaving infected houses or entering a town from plague-infested regions. Elizabeth I moved to Windsor during the summer of 1563 and forbade all travelers and goods from London from entering Windsor; she built gallows to back up her threat. In some local areas, other measures were enacted to attempt to control the spread of the plague. In London, houses visited by the plague were marked with a blue cross. In St. Margaret Westminster (130), dogs were killed in an attempt to control the spread of the plague. Although a few parishes along the Thames were severely affected, the geographic focus of this plague and later epidemics was nowhere near the locus of shipping and economic hubs, thus casting doubt on the idea that a major factor in plague outbreaks was the importation of infected rats and their fleas in goods arriving from Amsterdam or other foreign ports.

— 1578 —

Between the epidemic of 1563 and the rise in plague deaths in the late 1570s, the plague was sporadically active in many London parishes. In 1575, plague deaths were recorded in parishes as diverse and as far apart as St Olave Hart Street (85), St. Martin-in-the-Fields (128), and St. Margaret Westminster (130). However, Slack notes that 1578 was the crisis year in an epidemic that began in 1577 and continued until 1583; some parishes were affected by plague in 1577 while some did not become infected for several years. In 1579, St. Olave Hart Street parish (85) was one of the hardest hit by the plague. The plague epidemic of 1578 was relatively minor in terms of the total death toll, but it was unusually protracted, which makes it difficult to determine its effect on crisis mortality levels.

Although it was a relatively minor epidemic, the increasing number of plague deaths provoked public response during the epidemic and it was during this epidemic that the first national English rules were written to deal with plague outbreaks. In 1577, a set of questions was drafted about what had been done during the earlier London plague epidemics. The questions asked in this royal inquiry include a request for information on the number of people and houses affected and the number of people who died, as well as information on which parishes were involved. The query also sought information on what measures had been used to segregate the well from the ill, and to assure that infected shops and houses were shut up, as well as to find out what was done to provide for the impoverished. Finally, the questions asked if the Orders had been effectively enforced so that transgressors were suitably punished, and whether the Orders had been equitably applied to poor people and those who were better off.

Despite being a relatively minor outbreak, the appearance of the plague in 1577 it provoked official government concern. Since plague was almost always present within London, it is unknown if specific events triggered heightened official response to a few plague deaths. All plays were canceled from August 1 through Michaelmas, September 29, and the Mayor was chastised for not insuring that infected houses were discovered and shut up. Nonetheless, it was not until November 1, after mortality levels had begun to fall, that concern about plague finally led the Crown to instruct the Mayor "to close all such Innes, taverns, and ale-houses as are knowen to have been infected since Michaelmas laste." This suggests that Crown authorities did not think City officials moved more quickly enough to enforce the plague orders, which limit access to ale-houses in general and to close any that were the site of infection. The orders even decreed that infected inns must remove their signs for the duration of their quarantine.

In 1578, the year after the Privy Council's query was issued about the success of past anti-plague measures, national plague orders were enacted and published. These plague orders, which include 17 points, were printed along with advice from physicians on how to avoid, prevent and treat plague, and they were reprinted virtually verbatim several times over the succeeding decades until plague disappeared from England.

These 17 measures provided a broad range of instructions to public officials on how they should combat plague within as well as beyond City Liberties. They included instructions on quarantining houses and providing for their inhabitants, and on how to collect information about the number of deaths caused by the plague. They also included instructions on counting the number of infected people. The Orders also required that officials provide a special burial ground for those dead of the plague. Measure eight required that inexpensive preventative medicines be generally available during periods of pestilence. Measure 13 placed restrictions on the distribution of goods, such as bedding and clothes, that had belonged to plague victims, and required that these goods be either burnt or cleansed and aired before being used again. Measure 16 specifies that anyone, especially clergy members who claim that these orders contravene rules of Christian charity, will be restrained from speaking or preaching. The final measure, item 17 states that the contagion flourishes when people are disorderly and thus, the Queen's plague orders, which are an attempt to provide order and direction, will control the disease.

In the years following 1578, the plague epidemic

seems to have smoldered along for several years. In 1579, there were a few deaths reported and the plague was more or less evenly distributed from January through September. The next resurgence, which lasted only three months, began in August of 1581 and was followed by a period of endemic plague which again flared into an epidemic in August of 1582, which lasted through December of 1582. Through 1585, the plague continued to be endemic, although with low mortality levels, and then it receded in significance until late in 1592.

— 1593 —

After a respite from the last phase of the protracted 1578 epidemic, the first notice of increasing plague deaths appeared on August 13, 1592, when the plague was described as "dailie increasing in London." In September 1592, the Thame Fair was postponed and in October, the Lord Mayor was instructed to abandon plans for "the ceremonies of his appointment" and to use the money saved by this restraint to care for those whose houses were infected. Plague deaths continued to be reported throughout the winter of 1592-1593. In January of 1593, the Mayor restricted public entertainments within London as well as the entire region within seven miles of the city. In January, the Acts of the Privy Council noted that plague was on the upswing, but the first burial labeled as a plague in a parish register occurred on April 23, in St. Olave Hart Street Parish (85), which was one of the parishes that suffered most in the previous epidemic. By May, it was noted in The Calendar of State Papers that "the plague is very hot in London and other places of the realm so that great mortality is expected this summer." By June, as the epidemic continued to expand and the death tolls increased, all the public markets in the areas surrounding London were canceled.

London's mayor wanted to proceed with the Bartholomew Fair because of its economic importance, but his request was rebuked. The Privy Council recommended that his energy would be put to better use making sure that the crosses used to indicate plague-infected houses remained on the houses rather than being washed off. This admonishment to London's Mayor from the Privy Council indicates that controlling plague in London was viewed as an important step in controlling plague throughout the country.

In addition to preventive medicines and treatments, Kellwaye's Chapter 11, "Teacheth what orders magistrates and rulers of Citties and townes should cause to be observed," makes recommendations on civic controls. It has a similar set of prescriptions to those contained in the plague orders, although they emphasize different aspects of civil responsibility. Kellwaye emphasizes civic involvement that focuses on issues about the cleanliness of both the streets and fiber goods from infected houses, on assuring on the availability of good wholesome food and removing loose animals from the streets. The plague orders emphasize the need to raise taxes to enable infected houses to be closed and to provide the people within the closed houses with necessities. Kellwaye's eighth suggestion specifies that innkeepers should clean their stables of dung and filth every day because keeping it in the house for a week or two as people normally did, created "such stinking savor and unwholesome smell, as is able to infect the whole streete where it is," when it is moved. Kellwaye's advice serves not only to illustrate the perceived connection between bad odors and infections, it also serves to illustrate how noisome was the human environment of London in the late sixteenth century. His description also begs the question, if the smell of moving the dunghills through the street was so dangerous in the street, how hazardous could the stench have been when the dung remained inside the stables? The national Plague Orders place less emphasis on cleaning, but they do acknowledge the need to clean or 'aire' all material goods from infected homes before they are allowed to be used again, either by recovered victims or by others if the plague victims died.

Because the years between the epidemics of 1563 and 1603 seem to have been the period when rich and poor parishes became increasingly distinguished by differences in the intensity of their plague experiences, this 40-year period should be more carefully examined to discover changes that may explain how this distinction came to be exacerbated. Information from a few city center parishes should be examined carefully from the 1550s through 1600 to determine if parishes became increasingly segregated or if there were changes in the human environment that might have affected these changes.

This epidemic peaked in the third week of August 1593, earlier in the year than many other London plague epidemics. Neither Slack nor Shrewsbury provides quantitative summaries of London's parishes' plague experience during this epidemic. Shrewsbury says the epidemic was most devastating in "new housing-estates" and in slums around the docks, however, a map (Figure 2) made using his data does not completely support this argument. Creighton reports that the area of the Fleet was most devastated by the epidemic, but it was the parish of St. Katherine by the Tower seems to have experienced the highest mortalities, with 17 times its normal number of annual deaths.

Mapping some of the parishes that had mortality increases of from the high of 17 times normal to two parishes that suffered mortality levels only twice their normal levels shows that serious mortality was spread across the city from east of the Tower walls, to just beyond Westminster in the West. Nonetheless, except three suburban parishes, these severely hit parishes are clustered centrally in a north-south alignment within London. The cluster of badly hit parishes in 1593 shares some features with the cluster of worst-hit parishes in 1563; both clusters are within the walls and west of the Walbrook and both clusters show more overlap with parishes that Paul Slack lists as wealthy than to those he records as poorest. In one parish, St. Dunstan in the East (28), the effect of the two epidemics was quite different. In 1563, it was one of the least devastated parishes while in 1593, it was the second hardest hit with mortality levels.

Figure 2. London Plague 1593 Parish divisions from Justin Champion. Mortality level information from J.D. F. Shrewsbury.

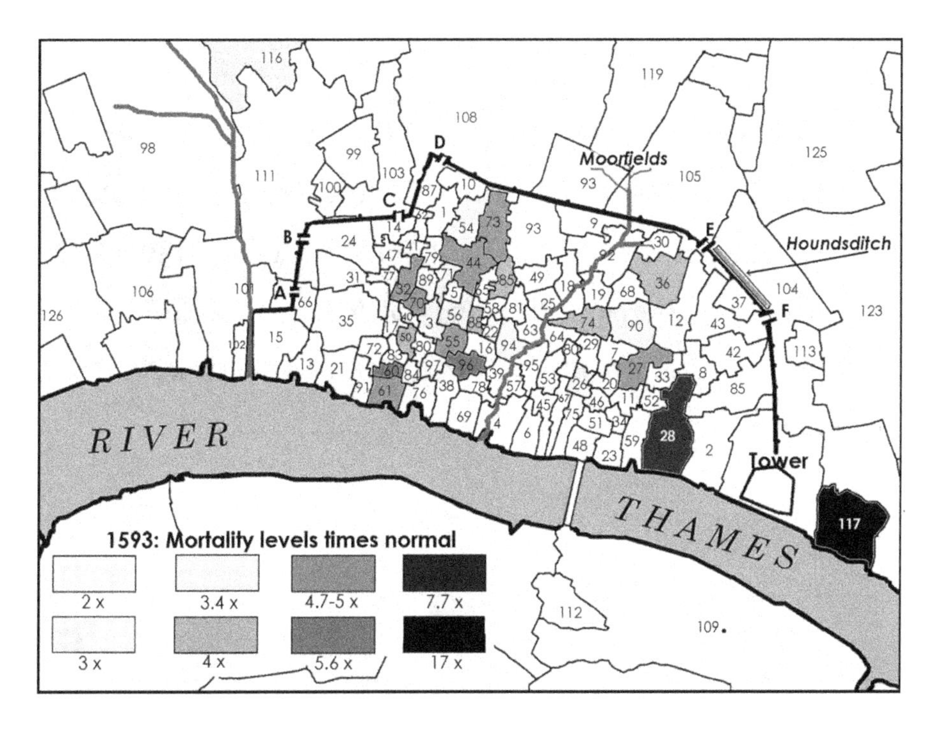

A) Ludgate, B) Newgate C) Aldersgate, D) Cripplegate E) Bishopsgate F) Aldgate

2x All Hallows Bread Street (3), St Mary Magdalen Milk Street (71)

3x St Peter Cornhill (90), St Martin in the Fields (128), St Mary le Bow Cheapside (56), St Stephen Walbrook (94), St Mary Aldermanbury (54), St Clement Eastcheap (26)

3.4x St. James Clerkenwell (116)

4x St Michael Cornhill (74), St Margret Moses (50), St Helen Bishopgate (36) St Olave Hart Street (85), St Pancras Soper Lane (88)

4.7x St Mary Somerst (61),

5x St Denis Backchurch (27), St Mary Alder Mary (55), St Vedast (32), St Matthew Friday Street (70), St Lawrence Jewry (44), St Michael Bassishaw (73) **x** St Thomas the Apostle (96), St Mary Mounthaw (60)

7**.7x** St Dunstan in the East, **17x** St Katharine by the Tower

— 1603 —

Only ten years after London was hit by the epidemic of 1593, it was visited by yet another devastating plague epidemic. 1603 was a traumatic year for London because it also saw the death of Elizabeth and the entry of her replacement by the Scottish King, James I. In early March, the plague was reported in Southwark but following this report, concern about the plague was overshadowed first by news of the Queen's death March 24 and then by planning for her funeral, which did not take place till April 28.

James I did not arrive in London until May 7. Plague is recorded on the London Bill of Mortality dated the first week of May. This plague epidemic was seen as a bad omen for the reign of the new King and it provided a serious test of his power and administration. It appears that neither London nor the crown issued comprehensive plague orders during this epidemic. In May, the Venetian Secretary in England sent a report home to Venice, stating that plague had killed people in nine parishes, that dread was rising, and that nothing had yet been done to control plague except mark infected houses and kill dogs. Records from St. Margaret Westminster (130), indicate that dog catchers killed 502 dogs. Although St. Margaret Westminster was singled out because of its proximity to the Royal Court, presumably, similar actions were undertaken in other parishes. On May 29, all gentlemen who were not needed to run the city or attend the court were ordered out of the city. In July, the St. James Fair and the King's coronation were postponed, and finally, on August 8, all fairs within a radius of 50 miles of London were canceled and ultimately the King decided not to enter London until winter. The Venetian Secretary reported that further restrictive edicts were issued in September when it was ordered that no one was to leave London. Not surprisingly, these orders seem to have done nothing to diminish the plague within London. The epidemic continued to rage through October and did not substantially abate until December.

Because plague was associated with overcrowding and "pestering" people within tenements, another proclamation was issued in

September, to limit and control tenements. No new tenants were to be admitted into infected houses "until it shall be thought safe [and] none of the rooms are to be pestered with multitudes of dwellers." The ultimate goal was to reduce the number of tenements within London, so it further stipulated that as tenements were destroyed, they were not to be replaced.

The congruence of the plague with the death of the sovereign affected contemporaneous perceptions and accounts. Thomas Dekker's account of the epidemic of 1603, entitled The Wonderful Yeare, makes it very clear that Queen Elizabeth's death "(like a thunder-clap) was able to kill thousands, it tooke away hearts from millions" and the requisite change of sovereign contributed to the dire consequences of this epidemic, and that equally, this change seems to have handicapped the crown's ability to effectively control the plague outbreak.

Figure 3 Plague in London 1603

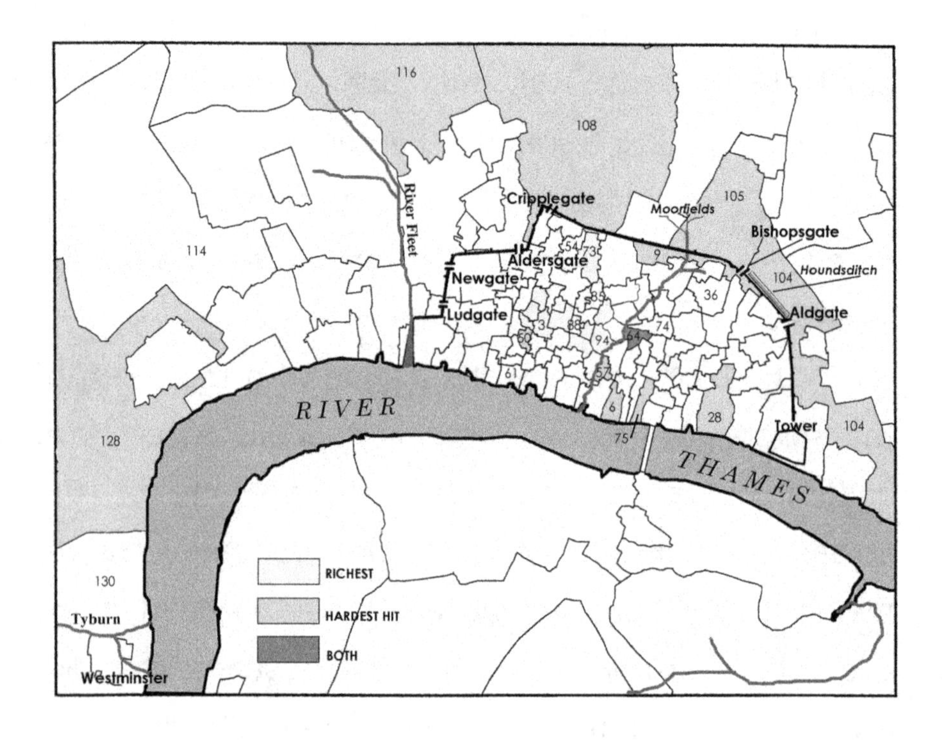

In this epidemic, the central and richer parishes suffered lower mortality rates than those at the periphery. According to Paul Slack among the ten hardest hit parishes were three large parishes just beyond the northern walls, St. Botolph Bishopsgate (105), St. Botolph Aldgate (104), and St. Giles Cripplegate (108), All Hallows on the Wall (9) just inside the walls from St. Botolph Bishopsgate, and three parishes along with Themes, St Dunstan in the East (28), St Michael Crooked Lane (75) and All Hallows the Less (6). Unlike in 1593, only three of the worst-hit parishes were centrally located, they were St Mary Woolnoth (64), St Mary Bothaw (57), and St Margaret Moses (50). Additionally, two distant parishes within the Liberties, St. James Clerkenwell (116) and St. Martin in the Fields (128) were particularly hard hit. This epidemic seems to have been widely distributed throughout the region around London (see figure 3).

In the years following the epidemic of 1603, the plague gradually diminished. By examining records from nine central London parishes – St. Michael Cornhill (74), St. Helen Bishopgate (36), All Hallows

Bread Street (3), St. Pancras Soper Lane (88), St. Mary Somerset (61), St. Michael Bassishaw (73), St. Stephen Walbrook (94), St. Olave Hart Street (85) and St. Mary Aldermanbury (54) – it is apparent that the number of reported plague deaths was lower in 1610 than in 1606, and that the percentage of total deaths reported as plague diminished slowly over the period. This pattern of gradual decline in mortality levels differs markedly from mortality patterns displayed during epidemics of highly infectious diseases like smallpox, which burn out quickly as the susceptible people are infected and become resistant to the disease. In 1608, the mayor and aldermen of London issued plague orders at the direction of the Privy Council. These orders, in addition to mandating that plague-visited houses be shut up and marked, regulated that cloth from these houses is cleaned and aired before being reused and that the streets be cleaned. Fleas, though were not linked to plague, were seen as a nuisance. In 1610, sixpence was paid for salt to be used to kill fleas in the pews of Westminster. London churchwardens also paid ten pence for an insecticidal powder to put in the Church linens.

Although these measures could only have been slightly effective, their use demonstrates that people were actively attempting to reduce their contact with fleas and perhaps suggests that fleas were quite abundant. The cumulative effect of these kinds of measures in conjunction with measures that mandated streets cleansing may have been enough to alter the rat and flea population and thus to affect how and where humans experienced plague.

— 1625 —

In the period between 1613 and 1624 London was free of epidemic plague, although it caused sporadic deaths. In 1624, at least some London parishes were troubled by many different diseases that produced increases in mortality levels.

Records from St. Katharine by the Wall (117), located outside the wall to the east of the Tower of London, suggest that by 1621 mortalities had already risen substantially and remained well above the 100 deaths a year that became typical after the plague epidemic of 1625. During the 1593 epidemic, deaths in St. Katharine by the Wall increased from about 40 a year to more than 650 deaths, while in 1625 the death total was approximately 225, up from a little over 100 deaths in the non-plague year 1620.

The increase in the normal death level in St Katharine by the Wall indicates that the population of that parish had increased substantially. Although the London epidemic of 1625 is generally considered one of the most serious plague epidemics St. Katharine by the Wall did not suffer as badly in 1625 as it had in

1593. Even during epidemics, the plague was not a universal experience across London.

During this epidemic, many of those who could flee did, although their departures may have been delayed by ceremonies necessitated by the King James' death and the arrival of the new Queen. Indeed, the Tuscan resident noted in correspondence that he thought that the real plague mortality figures were being hidden so that people would not flee London. Nonetheless, as Reverend Joseph Mead wrote in September,

> *The want and misery are the greatest here that ever living man knew; no trading at all; the rich are all gone; housekeepers and apprentices of manual trades begging in the streets, and that in such a lamentable manner as will make the strongest heart to yearn.*

The epidemic of 1625 like that of 1603 occurred in the year of a monarch's death and both years, affairs of state took precedence over attempts to combat the plague. Conduct of state affairs was protracted. James I died March 27, and his funeral was not until May 7. After James I's funeral, additional ceremonies related to the arrival of the new Queen, Henrietta Maria, and the coronation of Charles I, distracted official attention from the plague even longer. Visitors and nobility were encouraged to remain in London to participate in these ceremonies. Thus, during a long period, political attention was focused on the required pomp and ceremony, and away from increasing plague mortalities. Thus, laws were not enacted to restrict the spread of plague and in April, the number of parishes reporting plague increased from four to ten. By May the increase in plague deaths was making people nervous although it was not until June that plague was widely distributed across London. Despite the increasing mortalities, the Queen was brought to London on June 16. Ultimately as mortalities increased the court left London and King Charles' coronation was substantially postponed until October.

Parliament convened June 18, but in July it adjourned to Oxford. On Saturday, July 2 concern about plague prompted the declaration of a public fast accompanied by long religious services. By the end of July, a day of fasting and praying became a weekly event celebrated Wednesday. These attempts to stop the spread of plague failed and by the end of July, burials in London exceeded 3,500 a week, up from 305 a week in late April. In late August the death toll peaked, at over 5,000 deaths in one week.

After this peek weekly death numbers began to decrease and Charles I's coronation was finally held October after the number of plague deaths had considerably decreased. By November the death toll was back down to 319 in a week, which was still high according to the Tuscan Resident in London who had written that the average weekly mortality varied 130 to 180.

During the peak of the epidemic, attempts were made to assure that plague was not spread beyond London. Communication with London was heavily restricted. No one was allowed to enter the city, although letters could be sent into London with a payment, which indicates that regulations were not being followed. Boats were not allowed out of the Kingston docks and neither were people were permitted to leave London. To further protect the King and Court, while Charles was at Woodstock, no one from Woodstock was allowed to leave and return upon pain of death. Later, when the King moved to Read, similar rules were enacted to protect Reading. People living within three miles of Reading were prohibited from receiving goods from London. Although plague deaths had begun to decline by October, it was not until December 30 that all restrictions were lifted on travel and merchandise moving into and out of London.

In early October, when the number of deaths had begun to recede, and people began returning to London. Because of fear that plague would rebound with the influx of new potential victims, the government issued prohibitions against people entering London unless they were residents. Although London was under pressure to control the plague, evidence suggests that much of the official national concern about the plague was focused on protecting the King and his court from this epidemic. Nonetheless, plague orders issued by both the Crown and the City of London addressed the broader issue of what should be done in a time of plague. Two different but virtually identical Crown documents were printed; one was printed by John Bill and the other by Bonham Norton and John Bill and both are a reissued version of the plague orders set forth by Elizabeth I.

In 1625 previously issued London and Royal orders were revived and reissued. These emphasized the need to identify the ill, to provide for them, and to limit the movements. Also, the orders restrict the movement of the people such as searchers, caregivers, or watchmen who were required to be in contact with the ill. The orders clearly state that the ill and those who were exposed to plague-infected people should keep their distance from healthy people. As in the earlier epidemics, London officials were under considerable pressure to control the spread of the plague. The orders also placed considerable emphasis on keeping the London environment as clean as possible. They mandated that the streets should be cleaned daily and that plague-infected clothing should be well aired before being used again.

The 1625 London plague orders were a much simpler document than those issued by the crown.

They are only two-pages albeit very long pages long and look as if they may have been intended to be publicly posted. Nonetheless, despite their shorter length, they are very similar in content to those issued by the crown. Both sets emphasized a variety of techniques for arresting the spread of the plague. First, information was to be gathered about the location of the plague, and then the prescribed measures were to be taken. One of the primary tactics required was to isolate sick people from healthy. Doing so necessitated determining what houses were infected and then shutting infected houses up and restricting the liberty of those who dealt with the ill to mingle in society. Both infected houses and potentially infected people, searchers and surgeons, were required to be clearly labeled, thus allowing healthy people to steer clear of them. The London orders also required streets to be wet down, where wells were available, and swept twice a day, in addition to being raked daily except on Sundays.

The London orders also mandated the killing of dogs found roaming the streets. These orders were intended to apply to throughout the City Liberties including a region beyond the walls. Nonetheless, at the end of July, John Gore London's mayor assured the Privy Council that the fact that the plague had been more destructive outside the walls than within, was evidence of the efficacy of city measures. This statement suggests that the plague orders were better enforced and implemented within the walls than in the areas beyond the walls. It is possible that measures mandating cleanliness were effective at controlling the spread of the plague, where they were enforced. It also provides some evidence that the growing distinction between plague severity within and without the City Liberties could have been the result of differences in the enforcement of plague orders, rather than merely the result of more people living in the cramped squalid tenements of the suburbs.

Contemporaneous observers not only linked plague with overcrowded conditions, but they also recognized that it peaked during the heat of the year. In June a commentator wrote:

> *And that which makes us the more afraid is, that the sickness increaseth so fast, when we have*

> *had for a month together the extremest cold weather ever I knew in this season. What are we then to look for when the heats come on, and fruits grow ripe?*

The epidemic peaked in late summer, and by late November and early December, when cold weather arrived in London, plague mortalities throughout the greater London area virtually disappeared.

– 1636 –

Sporadic plague deaths were reported in the years following 1625. In March of 1629 fear of plague increased as "divers houses infected with plague in the parishes of St. Giles-in the Fields [114], St Leonard Shoreditch [119], and St Mary Whitechapel [123]" were reported and so previous plague orders were again reissued. Following these small epidemics of 1629 and 1630, pressure built to provide London with facilities to treat plague victims. As the number of plague victims rose, Orders were reissued that houses should be cleaned and the streets washed down daily, that no public meetings were to be held and that the shopkeepers should not keep their fruit stores in their house, but only in Thames street warehouses. Until 1636, however, only very few plague deaths were reported. The first plague death registered within the city was May 5 in the parish of St Peter Cornhill (90) although there had been earlier plague deaths registered in the out parishes, Stepney (125) and St Mary Whitechapel (123) in April and on May 4 a death in Bishopsgate and Aldgate was mentioned. Deaths continued to increase until September when they peaked, and then they gradually declined through December.

In 1636, many parishes on the outskirts of the city again saw increases in reported plague deaths: St Giles in the Fields (114) Giles Cripplegate (108), St Sepulchre (111), St James Clerkenwell (116), St Mary Islington (121), St. Katharine by the Tower, (117) Stepney (125), St Mary Whitechapel (123), St Leonard Shoreditch (119) and Isleworth. Although this epidemic was relatively mild and centered in the suburban parishes Shrewsbury notes that the Privy Council was quite concerned about the extent of the epidemic. The Privy Council again issued complaints to the Mayor that infected houses were not well marked. The Privy Council complained that the crosses and signs marking infected houses were placed so that they were not visible from the street, that many houses appeared to be unguarded and worse that people could be seen sitting outside, in front of infected houses. The Crown's plague orders issued in 1636 form an extensive publication. Although these orders include a reissue of the 17 regulations on how to control plague published in 1593, they also include several extensive orders for controlling rogues and vagabonds as well as regulations and reprints from court cases involving the division of buildings into tenements. Thus, the plague orders of 1636 seem to indicate that maintaining jurisdiction over disorderly behavior was considered central to the control of the spread of the plague. Emphasis on controlling disorderly behavior may have been a response to the location of plague hot spots or it may have been affected by larger political disturbances.

During the years between 1636 and 1665, occasional plague outbreaks caused deaths, but none was as serious as the epidemic of 1635-36. London issued expanded plague orders that continued to require street cleaning and fumigating of infected houses by burning various sorts of smelly herbs. These orders differed from earlier ones primarily in that they required houses to be shut up for 28, rather than 40 days after the last resident had contracted the plague. There were a few outbreaks between 1640 - 1647. During the Interregnum, regulations concerning the division of buildings into tenements, and construction of new tenements in London's open spaces were relaxed or at least less-strenuously enforced. After 1647, the number of deaths registered as plague returned to the normal endemic pattern fluctuating between a couple, and less than a few hundred annually, until 1665.

– 1665 –

In 1665 London experienced what is the most famous and most studied, and best documented of all its plague epidemics. Although the first deaths of this final great London epidemic are usually considered to have been those reported in St. Giles in the Fields (114), during the fall of 1664, plague deaths were also registered in Westminster. Nonetheless, plague first reached epidemic proportions in St. Giles in the Fields, and the records of these deaths have come to take precedence. The plague seems to have begun very slowly in late 1664, as the following contemporaneous report explains:

> *And being constrained to a house or two, the seeds of it confined themselves to a hard-frosty winter of near continuance: it lay sleep from Christmas to the middle of February, and then broke out again in the same parish; and after another long rest til April, put forth the malignant quality again as soon as the warmth*

of spring gave sufficient force, and the distemper showed itself in the same place. At the beginning it took one here, and another half a mile off; then appeared again where it was first: neither can it be proved that these ever met; especially after houses were shut up.

The records of St. Paul Convent Garden (127) lists the burial on April 12, Margaret, daughter of John Ponteus as a plague death, followed by another plague death in May. Neither of these was included in the plague deaths listed in the Bills of Mortality. In the last full week in April, two other plague deaths were reported in nearby St. Giles in the Fields. No plague deaths were reported in the next week, but in the first full week of May, several plague deaths were recorded in the Bills of Mortality: three in St. Giles in the Fields and four in close by St. Clement Danes (126). Also, two other parishes reported one plague death each in the first May: St. Andrew Holborn (98), a parish partially within and without City Liberties, and finally

St. Mary Woolchurch (63), located right in the center of London within the wall to report a plague death (see figure 4).

Figure 4. London Plague 1665

Shaded parishes experienced the first plague deaths.

Numbered parishes are those mentioned in the text.

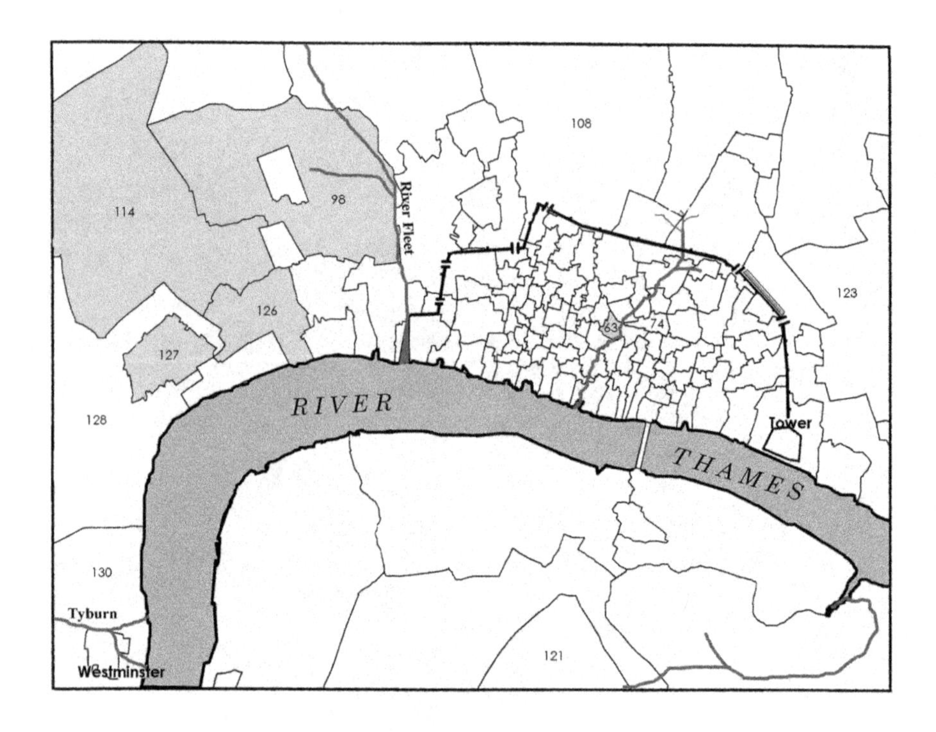

Despite these few plague deaths in April, the Privy Council required that houses suspected of infection be inspected, and in May constables were directed to make sure that infected houses were promptly shut up. Shortly after this, the "the justices were authorized by warrant to purchase ground for the erection of pest-houses" and to provide access to the newly constructed pest-houses. Bell argues that these proactive attempts to battle plague would not have been undertaken if those in charge had not known that there had been more plague deaths than had been reported in the Bills of Mortality.

By May 12, a Privy Council Committee had been appointed to devise a plan to control the plague. After considering it for one week, committee members made their first report, which recommended building another pest-house in St. Giles in the Fields, restricting the number of alehouses in infected areas, and strictly enforcing the previously enacted laws against 'inmates' or lodgers. London's mayor issued a proclamation requiring Londoners to fulfill their civic duty and clean the streets in front of their houses. Two weeks later, he complained that the streets were dirtier than ever and followed this up by requiring aldermen to inspect their wards and report infractions. Although the requirement that aldermen check their wards was mandated in previous plague orders, it was not stipulated in those printed in 1665, which simply state that the streets are to be cleaned and that the aldermen shall meet at least weekly to discuss the implementation of the orders.

In an additional attempt to control the spread of plague, efforts were made to restrict travel throughout greater London. A path behind Christ's Hospital that ran along the old town ditch was closed because the plague was associated with the filth there.

Magistrates also attempted to restrict access into and out of St. Giles in the Fields by placing guards around the parish. Despite these measures, mortality continued to increase. Until August, the bulk of the deaths occurred in St Giles in the Fields; however, plague deaths were also being reported more widely, for example, east of the wall in St Mary Whitechapel (123) and even in St Mary Islington (121) on the south side of the Thames. The next to be hit hard by plague was St Giles Cripplegate (108) north of the wall, which had a population one hundred times the population of any single parish within the walls of London.

Variation in the population of parishes makes comparing mortality figures virtually worthless, in terms of determining relative mortality levels. Notwithstanding, parish death figures do provide some clues to the enormous stresses faced by some of the outlying parishes. All the burials and funerals recorded in a parish were handled by the officials of a single church. Additionally, each parish was largely responsible for financially supporting indigent plague victims, delivering their food, and paying for the watchmen who ensured that their closed houses remained closed. Thus, the finances and social and physical infrastructure of these parishes with huge mortality figures were severely tested. According to Walter George Bell, in these outlying regions plague spread along the main arteries. In St. Giles in the Fields (114), it spread along Dury Lane and Holborn, and in St. Martin in the-Fields (128) it spread along Long Acre. Bell provides no concrete evidence for his statement, that plague spread along the main thoroughfares' arteries, rather than more haphazardly, however, because outlying parishes were heavily built up along main arteries it seems logical. In the suburbs the built-up areas backed on too much more rural areas, so one would expect a disease carried by commensal rats and fleas to be spread through the interconnected buildings rather than across streets or fields. Within London's walls, plague distribution may have been more erratic because rat populations were separated by a more complicated pattern of streets and interconnected buildings.

Samuel Pepys (1633-1703) first mentions the epidemic June 7, when he wrote,

> *This day, much against my will, I did in Drury Lane see two or three houses marked with a red cross upon the doors, and 'Lord have mercy upon us' writ there; which was a sad sight to me, being the first of the kind that, to my remembrance, I ever saw. It put me into an ill conception of myself*

and my smell, so that I was forced to buy some roll-tobacco to smell to and chaw, which took away the apprehension.

In mid-June, the Barnwell Fair was canceled, and on June 21 when Pepys was at Cripplegate he saw that people traveling by the coach and wagon load were heading out of town. Nonetheless, even in the first week of July, there were still parishes within the city walls that had yet to report any plague deaths. One of these fortunate parishes, was St. Michael Cornhill (74), despite being one of the central market parishes, at the crossroads of several of the main streets through London. In the epidemic of 1593, it had been one of the parish's hard-hit, although in the subsequent epidemics St. Michael Cornhill had also experienced a relatively low level of endemic plague.

It has been generally accepted that the 1665 plague epidemic was introduced to London from Amsterdam because the plague had been virtually absent from London, while it had been rampant in Amsterdam. It is possible that plague was introduced from Amsterdam into London either by rats or fleas; however, the pattern of its spread within London, beginning as this epidemic did in a northwestern suburb, suggests that enzootic plague within London became epidemic for a reason other than direct importation from abroad. However, Dr. Thomas Cocke assumes that the plague entered St Giles in the Fields in a bundle of fur from Holland, while Dr. Nathaniel Hodges traced the outbreak of plague to packages of Turkish "bailes of cotton or silk, which is a strange preserver of the Pestilential Steams" imported via Holland. William Boghurst (c.1630 –1685), however, noticed that a few cases of plague had occurred in several of London's outer parishes over a several-year period.

It is unlikely that this epidemic was introduced from Amsterdam, not only because the plague first broke out far from the port area but also because in 1664 ships from Amsterdam had had restrictions placed on them because of the plague. Ships arriving in London from Amsterdam were required to wait 30 days before either people or merchandise were allowed to be unloaded. As the plague spread throughout the Netherlands these restrictions were expanded, first to include ships from Rotterdam and then Zeeland, and the restrictions were lengthened to 40 days. According to Moote and Moote, when it came to a shipment of horses for King Charles II, the regulations were not followed, and they further suggest that shipments for the King's Navy were not subject to the rules.

Only after the plague rumbled for several months in the extramural parishes of London was the first plague burial recorded within the walls of London. That first recorded plague death was in centrally located St Mary Woolchurch (63). Defoe claims that this first death within walls of the city was the result of flight from St Giles in the Fields by someone who did not realize that he was already infected. This has come to be the accepted explanation for how plague entered the city within the walls.

London's dreaded visitation of 1665 killed more people than any other single London epidemic. At least 68,747 died of the plague that year, double the number of people killed in any one of the previous epidemics. Though these official numbers seem huge, many contemporaneous observers and subsequent commentators and historians have believed that the official reports seriously undercounted deaths. Nonetheless, the epidemic was well documented by its contemporaries, not only by official record keepers but also by doctors and prominent figures who kept diaries and wrote letters into and out of London, many of which have survived. Also, despite the fire of 1666 and the subsequent reorganizing of several parishes, many of the parish registers, which were kept amazingly up to date despite the hectic nature of the plague year, survived and have been thoroughly studied. Finally, other factors such as hearth tax records have been combined with plague death records to create a fairly detailed picture of the wealth of each parish and the number of households it contained.

Various people made notes about the weather they associated with this plague epidemic, but the weather patterns they remembered do not always correspond well with the actual weather, however, Bell and Harvey reported that the winter of 1664 was cold and evidence suggests that it was. On the other hand, the summer of 1665 seems to have been almost as warm as 1664 and a little cooler than 1666 which was an abnormally dry year in London. Linking plague to specific weather and climate is difficult because, for example, good harvests may produce more rats, but on the other hand, bad weather would have driven the rats into closer contact with people.

Diaries including those of Pepys and John Evelyn say that many people who could afford to do so fled London as the bills of mortality reported increasing numbers of plague deaths. It has been suggested that one of the reasons that the wealthy suffered dramatically less during the epidemic was that they left London. However, the evidence from their diaries suggests that explanation is too simple. Although Evelyn left London during the Epidemic, he returned, and while Pepys, remained in London throughout most of the epidemic, he visited his wife frequently and spent some time in the country. Pepys moved his wife and most of his household out of London during the height of the plague, yet he frequently traveled to visit her and then returned to London. Despite the large numbers of people who fled London during the 1665 epidemic, the surrounding regions were only sporadically afflicted with the plague, which suggests that this flight and visits such as Pepys' did not do much to spread the plague. However, while the London epidemic had largely burnt itself out, by March 1666, in the surrounding regions there were plague mortalities reported as late October 1666. It is unclear the extent to which people fleeing London was responsible for the delayed plague outbreaks in the regions surrounding London. What is clear is that despite these lingering plague loci, the disease was not reintroduced back into London.

In Conclusion

The physical and social geography of London was substantially constant throughout London's 300 years of endemic plague, although metropolitan London did experience dramatic growth population that caused expansion far out into the suburban regions and increased its population density. Nonetheless, plague orders mandated a remarkably consistent set of measures to be used to combat the plague. Throughout the plague's final 100 years, people were told to clean the streets, clean fabrics and fumigate houses, kill dogs, and keep infected people and those exposed to people with plague segregated from the healthy population. Yet, despite the consistency in the methods recommended for combating plague, the locus of epidemics moved from the form of the central, relatively wealthy parishes, to the relatively poorer suburban parishes, during the final century of plague in London. Despite the observable changes in the focus of plague distribution, its spread from place to place remained erratic.

Londoners noticed that although everyone was

susceptible to plague, it seemed to be most devastating among the people who were poor and pestered, or crowded, into tenements in back alleys and the unregulated, squalid housing of the suburbs. Plague regulations attempted to control the poor and thus to control the spread of plague, however, evidence suggests that these orders were more effective within the wealthier parishes inside the City walls.

Literature Review and Argument for *Yersinia Pestis*

Chroniclers, physicians, and historians have been writing about the plagues of the Second Pandemic since even before the disease reached Europe. For the most part, documents written before 1722, when plague virtually disappeared from Western Europe, focused on methods of controlling, avoiding, preventing, and escaping the disease, as well as treating people who had become sick with it. The authors of the earliest treatises also attempted to understand the causes of the disease in the expectation that if the causes were understood, the epidemic could be avoided. Both early and later treatises frequently discussed plague as a significant challenge to basic social order and addressed means of mitigating its impact on society.

Though many contemporaneous plague treatises exist, twentieth and twenty-first-century plague researchers face many unsolved questions. Almost every aspect of the epidemic that ravaged Europe from 1347 to 1350 provides a point of contention, as do aspects of subsequent epidemics that sporadically returned throughout Europe for the next 300 or more years. It is generally accepted that the initial epidemic killed between one quarter and one half of the population of Europe. When the smaller secondary epidemics between 1350 and 1361 are considered in conjunction with the initial epidemic, the mortality figures mentioned are frequently substantially higher, often sixty percent or even greater. However, mortality figures mentioned, at least occasionally, range from an estimate as low as five percent by J. F. D. Shrewsbury, to estimates as high as seventy-five percent, at least regionally.

Researchers' positions are divided between several different viewpoints concerning the level of mortalities and the cause of the medieval and early modern plagues. Although many researchers think medieval descriptions of plague symptoms come close to matching modern descriptions, and the case mortality for the three forms of plague mesh well with the Modern Pandemic, the overall mortality levels during epidemic outbreaks of the Modern Pandemic have been much lower than outbreaks during the Second Pandemic. In general, researchers who accept the historical tradition of the early twentieth century and believe that *Y. pestis* was the causative agent of plague epidemics, and accept moderate mortality figures for the initial wave of the pandemic. On the other hand, researchers who think that plague epidemics were the result of a disease more contagious than bubonic plague tend to be more willing to accept evidence and chroniclers' reports of very high death rates associated with plague epidemics. Thus, because Philip Ziegler, Christopher Morris, and Paul Slack believe that *Y. pestis* was the primary causative agent in the plague epidemics, they accept relatively cautious mortality figures for the initial epidemic of between 30 and 45 percent mortality. J. D. F. Shrewsbury takes an even stronger stance. He argues that the overall mortality was greatly overstated because he believes that *Y. pestis* is incapable of producing the high mortalities that were reported, especially in rural areas. Because Shrewsbury cannot completely discount mortality reports, he attributes some of the mortality to something other than the plague, although he credits *Y. pestis* with a greater role in the initial epidemic of 1348-50 than in subsequent epidemics.

Susan Scott and Christopher J. Duncan, on the other hand, accept very high mortality figures, but deny entirely the claim that the initial epidemic was caused by *Y. pestis* and argue instead that these massive mortalities were the result of a form of hemorrhagic fever. Graham Twigg, unlike Shrewsbury, does not discount a high the overall level of mortality; but rather he attributes the bulk of the mortality to anthrax. Like Twigg, Samuel K. Cohn accepts high mortality figures, especially for the initial epidemic wave, and like Twigg, he does not believe *Y. pestis* was responsible for the high mortalities; however, unlike Twigg, Cohn does not concern himself with what disease was responsible for the plague. Ole J. Benedictow is one of the very few researchers who believe that the Second Pandemic was caused by *Y. pestis* and accepts the idea that it produced very high mortalities. To sum up, the research on plagues of medieval and early modern Europe has led to a few points of consensus. The principal point of consensus is that an enormous number of people died in a relatively short period, while everything else is a point of contention. Details of precisely how many people died and what diseases caused their deaths are hotly disputed.

Between 1985 and 2005 many works addressing plague, or the epidemic diseases of medieval and early modern Europe have been published. Because these new looks at this old epidemic are being fueled by current events, many of the new examinations of past plagues have focused on their biological nature, including both the microbial causes and disease vectors. Of course, in the past when interest in historic plagues did surface, it also was driven by contemporaneous concerns. The chaos surrounding World War I served to increase interest in the turmoil produced in 1348 by the first wave of the Second Pandemic. Current interest has been fueled by the realization that infectious diseases that were thought to have been conquered have not been. As McNeill noted in the introduction to the 1998 re-release of his book *Plagues and Peoples*, when the book was originally published "in 1976 many doctors believed that infectious diseases had lost their power to affect human lives seriously." Alfred W. Crosby in his preface to the new edition of *America's Forgotten Pandemic*, notes that in 1969 the Surgeon General of the United States declared "that we had left infectious diseases behind in our dust," but by the late 1980s, it was all too clear that this was a wholly inaccurate statement. In light of the many new biological threats, including bird flu, Severe Acute Respiratory Syndrome (SARs) and HIV-AIDs, and the growing concern over new and improved biological weapons, the Surgeon General's statement seem unbelievably naive. Given the rapid changes in perception about the dangers of contagious diseases, it is not surprising that interest in the great historic plague epidemics has grown over the past few decades. Given the impetus of late-twentieth-century plague research, it is not surprising that most historians have failed to examine how medieval society may have influenced the course of plague epidemics. This is an oversight this thesis has begun to redress.

After the *Yersinia pestis* bacterium was identified in Hong Kong in 1894, it was generally accepted that the disease it produces, plague, was the epidemic disease that devastated England, most of Europe, and much of Asia in the fourteenth, fifteenth and sixteenth centuries. However, since the mid-1980s, it has become increasingly common to question this theory. Recently, Cohn has gone so far as to claim that while he cannot suggest what the causative agent of the plague was, he is sure it was not *Y. pestis*. Other investigators argue that while *Y. pestis* may have played a role in the epidemics, other diseases must have accounted for a considerable proportion of the high mortality.

This upswing in interest in historic plague epidemics is the result of multiple kinds of inquiries into the events of the late twentieth century. Advances in the fields of microbiology and epidemiology have driven much of the recent research on the plague.

Until recently, researchers who were attempting to investigate epidemics of the Second Pandemic were limited to using information gained from current epidemics and attempting to apply it to the epidemics of the past. Scientists are now making use of new technologies that allow them to directly investigate the genetic materials from the microbes that caused historic epidemics. Genetic examination of humans with resistance to AIDS/HIV provides evidence that an epidemic 680 years ago selected for a genetic mutation, known as CCR5-delta, which now has a ten percent frequency among Northern Europeans. Naturally, the time frame of the selection for this mutation led researchers to propose that the epidemic disease that swept through Europe in 1347-50 selected for the CCR5-delta mutation. This and other evidence of the impact that diseases of the past might have had on human genetics of the present are providing a definite impetus to determine precisely the nature of the pandemic that swept Europe in 1348 and sporadically throughout subsequent centuries. Plus, interest in *Y. pestis* is being encouraged by its potential as a biological weapon.

Stimulated by new diseases and fueled by new technologies, as well as by the desire to discover the disease that may have selected for the CCR5-delta mutation, researchers have begun to seek DNA evidence that *Yersinia pestis* existed in medieval

Europe. Teams including Didier Raoult and Michel Drancourt have identified *Y. pestis* in dental pulp from several medieval and early modern gravesites that have been excavated in France. The bodies selected for examination were chosen from mass graves that were likely to have been created during plague epidemics, for example, the dental pulp examined was from three separate individuals found at Sainte-Damien in Montpellier. Because this cemetery was in use before and after the 1348 plague epidemic, the bodies selected for testing were chosen based on their location within the cemetery. The chosen bodies were located on top of a thirteenth-century rubble pile and behind a fourteenth-century wall, and, also, they were buried without shrouds suggesting they were buried during a mortality crisis. In addition to these medieval bodies, they tested bodies from cemeteries of the Justinian Pandemic era, as well as bodies that are presumed to have died as the result of epidemics of the early 1700s. The results of these tests indicate that *Y. pestis* was present. However, another team of researchers led by Tom Gilbert examined samples from other medieval bodies but found no evidence of

Y. pestis and claim that Drancourt and Raoult's methodology is faulty. Other researchers simply acknowledge that more than one disease was present during epidemic periods. Although the work is not conclusive, Drancourt and Raoult's results provide more evidence that *Y. pestis* was a factor in the Second Pandemic.

Nonetheless, for the foreseeable future, modern scientific analysis of human remains will serve to supplement but not replace, a historic analysis of what disease produced historic epidemics.

Scientific research affects the understanding of historic epidemics, and, conversely, historical understanding of diseases has influenced scientific research. The initial development of modern plague research, especially during outbreaks of the late nineteenth century, was influenced as much by an understanding of the past as by contemporary developments. Plague investigators of the late nineteenth and early twentieth century were influenced by their understanding of the earlier epidemics, especially the outbreak in 1348 and the final devastating outbreaks in London and Marseille.

When reports of the plague epidemic in Hong Kong reached Europe in the 1890s, Europeans rushed to Hong Kong and invested heavily in solving the mysteries of this modern epidemic because of fear that plague would again invade Europe producing mortalities similar to those of the past. Although before the late nineteenth century, the plagues of the late medieval period had not been extensively studied, the devastation caused by later plague epidemics, such as the 1665 Great Plague of London, was well known if only from fictionalized sources such as Defoe's *Journal of the Plague Year*. Researchers knew or thought they knew, that plague was deadly, terrifying, and extremely contagious, and this affected their investigations.

Of equal importance to the initial investigations into the modern plague epidemic was that plague in Hong Kong came to the attention of Europeans and that the *Y. pestis* bacterium itself was isolated, less than 20 years after the first microbe responsible for a disease, *Bacillus anthracis,* was isolated and identified. Thus, investigations of plague were influenced by a modern scientific understanding of contagion and how diseases can be expected to pass from person to person. Although researchers working in the field gradually came to realize that plague is not readily communicable from one person to another, this preconception, based on a scientific understanding combined with earlier descriptions of a terrible infectious disease, influenced the initial investigations.

Gradually researchers came to notice and to accept that the epidemiological patterns observed in the modern plague pandemic varied in several critical details from those noted in the public records and described by contemporaneous observers of the Second Pandemic. Increasingly, these differences are causing epidemiologists, microbiologists, and historians to question the true nature of the epidemics of the medieval and early modern periods. In the second plague pandemic's spread across Europe, the disease spread much faster than it does today, even without the benefit of truly rapid transportation. Nonetheless, mortality rates for the medieval plague and untreated plague cases in the modern pandemic are quite similar, although the morbidity rates of the two plague pandemics differ by several factors of ten.

Also, symptoms described by observers of the two pandemics vary greatly. The debate over the true nature of the earlier pandemics is heated and includes arguments based on different and conflicting contemporaneous historic descriptions and knowledge developed during twentieth-century research on plague and other diseases. Researchers on both sides of the argument accuse their opponents of being overly subjective in their selection of facts. Cohn argues that the discussion about the problem is made especially complicated because researchers are too biased by their views to examine the data dispassionately.

Many of the arguments about the nature of epidemics of the Second Pandemic revolve around the interpretation and significance of various bits of evidence. It is clear that the Second Pandemic spread across Europe much more rapidly than the Modern Pandemic, but it is less clear what this signifies: either medieval social conditions were very different or the Second Pandemic was the result of a different disease. Another issue open to various interpretations is what medieval and early modern commentators meant when they described the plague as very contagious. Cohn assumes that the early commentators meant something akin to our understanding of contagion, that pestilence was easily spread from one person to another. The evidence of the initial expansion of plague across Europe supports this idea, however, all subsequent outbreaks spread much more erratically. Thus, statements such as Guy de Chauliac's that plague was so contagious that "even by looking at one another people caught it," which were continually reiterated throughout the Second Pandemic can be interpreted as an indication that people found its spread uniquely inexplicable rather than as evidence that plague was especially contagious. Descriptions of the plague being spread by the glance of an eye can be taken to mean that plague was very contagious, however, these explanations could instead be taken to indicate that people did not know how the plague was spread from one person to another while passing over many in between. Kellwaye describes smallpox, a disease we now understand to be very contagious, as being a heritable disease, passed from mother to baby in menstrual blood, which can occasionally become manifest later in life, rather than in childhood.

Neither the argument for nor the argument against, *Y. pestis* as the primary cause of the devastating epidemics of the late medieval and early modern period is wholly conclusive, but at present the case for *Y. pestis* is more convincing than any other. The material that identifies *Y. pestis* as the primary biological agent responsible for the plague of the medieval and early modern periods includes evidence of both a biological or epidemiological nature and of a cultural or social nature. Until extensive DNA work can be done to positively identify the pathogens that are common to widely disparate mass burials of the epidemics of the medieval and early modern periods, I think that the most convincing evidence that plague was produced by *Y. pestis* is provided by an analysis of societal responses and an ecological understanding of plague. Unfortunately, one of the most telling societal responses to the plague provides only negative evidence. Despite the many years that plague afflicted Europe, there remained a general lack of understanding of how the disease was spread. This suggests that the disease had a complicated etiology and process of transmission, including multiple organic and behavioral components, and that all of these factors played a role. Therefore, it is only by considering all of these factors that the true epidemic patterns can be seen. People faced with the death and destruction caused by plagues had neither the emotional distance, financial wherewithal, scientific understanding, or knowledge to investigate how the plague spread.

Some of the early descriptions of how the disease that spread through Europe and across England presented in its victim's sound very much like modern clinical descriptions of the disease caused by *Y. pestis*, but some do not. One of the difficulties of arguing that many or a majority of the great mortalities of the late medieval and early modern period were one disease is that during the Second Pandemic, many people including physicians, surgeons, and social observers, did not pay attention to or to organize the specifics of disease symptoms and classifications in a modern scientific format. Thus, historic descriptions of diseases do not correspond well with modern descriptions. While it is true that science, as it is understood in the twenty-first century, was only in its infancy when plague disappeared from England, it is more accurate to say that during the medieval and early modern periods people were not concerned with the same classifications as modern scientists. Within the then-accepted humoral system of understanding disease and the body, it was easy to accept that disease, whether produced by miasma or toxin, would manifest differently in people with different humoral compositions. This does not mean, however, that diseases were not differentiated. In 1593, Kellwaye wrote that one of the warnings of an upcoming plague epidemic is "when the small Poxe doth generally abounde both in yong and olde people." Paul Slack notes two cases, one from the 1580s and one from 1626, in which a disease was described as being similar to the plague but not the plague. Kellwaye's contention that smallpox outbreaks precede or in some way signal the arrival of the plague is inexplicable, but his comment does indicate that diseases were differentiated and further it suggests that the plague was considered uniquely terrible.

The most often cited early0 description of plague is one by Boccaccio in the introduction of the *Decameron*. However, Guy de Chauliac also describes the two forms of plague, with different symptoms and mortality rates that the disease took. The two forms of the disease they described closely match two of the three forms of plague currently accepted as plague: bubonic and pneumonic. Cohn points out that contemporaneous descriptions of separate plague forms are rare outside of Italy. However, as it is highly unlikely that the disease that invaded the Italian peninsula in 1347 was not the same disease that spread throughout Europe, in 1348, Cohn's observation should be taken as a demonstration of the variability of reported observations rather than as evidence that northerners were afflicted by a different disease. Boccaccio also reported that the disease began in Florence with a variety of symptoms that were different from those that were reported in the East. That various observer reported on different symptoms for the first extensive serious epidemic in several hundred years doesn't necessarily indicate the presence of numerous diseases. It is much more likely that these divergent reports are all describing one disease, and that the descriptions vary because of different cultural preconceptions about which symptoms were most significant, possibly coupled with real differences in symptomology due to varying environmental factors. Evidence from subsequent outbreaks is less clear cut because outbreaks occurred sporadically across time and space without any clear evidence if, or how, the plague was spread from one location to the next.

That Boccaccio's description of the initial outbreak of the second pandemic in Florence closely matches a modern clinical description is important because of this *Y. pestis* epidemic did vary significantly from the modern experience of how this disease behaves. In addition to producing much higher morbidity rates, the initial wave of the medieval pandemic traveled across Europe much faster than the modern pandemic has traveled. The first epidemics of the modern pandemic seem to have begun in the Chinese province of Yunnan in the mid-1800s. Over time it has spread more widely than in earlier outbreaks and has now reached both the Americas and Japan. However, although the plague has now reached new regions of the world, it traveled considerably more slowly in the early 1900s than it did in the 1300s, despite the existence of the steam engine and the railroad and has spread no faster since the advent of the airplane. Although more widely dispersed, it has produced only localized epidemic outbreaks in both North America and Europe and it has had virtually no impact on England. The plague arrived in San Francisco in 1900. The plague has only very slowly dispersed into the surrounding countryside at the rate of about one mile a year. In contrast, the initial medieval plague epidemic spread at a rate closer to one mile a day. This extreme disparity between the speed of the spread of the modern pandemic that we know was caused by *Y. pestis,* and the speed of the earlier pandemic has led some historians and epidemiologists to postulate that the epidemics of the late medieval and early modern period must have been caused by some other biological agent. They have not, however, explored environmental factors that may explain differences in dispersal, such as changes in containers or modes of transportation that may have facilitated the movement of rats or fleas. The difference in the speed with which epidemics have spread may be related to environmental changes and changes in the ecology of the hosts and the behavior of the human victims.

The plague of the medieval and early modern periods seems to have occurred in the same three forms that are seen in plague outbreaks today, but the pneumonic and septicemic forms, which have substantially higher fatality rates than the bubonic form, apparently were much more common in the Second Pandemic than today. The pneumonic and septicemic forms of plague are dramatic and victims of these forms can die very rapidly. Death from the pneumonic form often takes as little as three days, while in the case of the septicemic form, a person may wake up well and yet be dead before nightfall. Because these forms are so terrifying and remarkable, it is entirely possible that deaths caused by these forms were overemphasized in reports in medieval and early modern epidemics. Conversely, these forms of plague may now be under-acknowledged since they are now considered rare and thus relatively insignificant, despite the dramatically rapid deaths they produce. Additionally, it has been assumed that the virtual absence of pneumonic plague explains the relatively low mortality in the Third or Modern Pandemic. Whether or not the pneumonic form of plague has been underreported, it is clear that during the Modern Pandemic the pneumonic form of plague has not produced the rapidly spreading epidemics or extremely high mortalities of the Second Pandemic.

Pneumonic plague has been a significant factor in the reoccurring plague epidemics in Madagascar, although plague mortality levels there have never reached the levels of the outbreaks in the Second Pandemic. The pneumonic form also played a role in fulminating an epidemic outbreak in Manchuria during the winter of 1910-11. It developed among people who lived in unventilated and crowded conditions and although it was spread as people fled by train, many people who rode in a carriage with pneumonic plague victims did not become ill. The epidemic produced the death of at least 50,000 and was the largest of the Modern Pandemic, but still produced only population mortality of about .04 percent. This suggests that the strenuous effort of flight on foot may have contributed to creating cases of pneumonic plague, and especially secondary pneumonic plague that was more significant in earlier pandemics. The pneumonic form erupted during an episode of the bubonic that was rampant among tarabagan hunters who had flocked to Manchuria from both Russia and China to collect skins of these marmots of the Gobi region because of exceptionally high fur prices. The hunters lived in extremely crowded unventilated rooms that were frequently at least partially underground, due to the cold conditions in Manchuria. With as many as 40 men living in a room 15 by 15 by 12 feet high, it is not surprising that pneumonic plague became predominant once the men became infected by fleas from the tarabagans they trapped. As the disease spread, panic spread along with it, and the visiting hunters attempted to flee Manchuria. Infected people who fled the epidemic on foot experienced pneumonic plague at high rates. This suggests that protracted exertion after infection with *Y. pestis* can greatly increase the likelihood that an infected person will develop, and then spread, the pneumonic form of the disease. Because the flight was a commonly reported response to plague during the Second Pandemic, the evidence from the Manchurian epidemic provides one possible explanation for how *Y. pestis* could have been spread so quickly during its initial dispersal across Europe in 1347-50. As physically stressed people caught the plague and fled their homes, their exertions increased their likelihood of developing pneumonic plague, which made them the actual carriers of the disease, independent of rats.

Although the Manchurian epidemic appears relatively insignificant when compared with medieval plagues, it was large compared with other epidemics of the Modern Pandemic. Unfortunately, it is impossible to compare the number of deaths in the Manchurian epidemic with those of the early individual epidemics in the Second Pandemic because the reports of the total number of dead in those epidemics are unreliable; for example, the total numbers of deaths reported for a given town often exceed the number of people who could conceivably have been living there. However, reports from the last century of the Second Pandemic are considerably more reliable.

The death tolls in London during several of the last major epidemics were 26,350 in 1625, 10,400 in 1636, and 55,797 in 1665. To put the Manchurian epidemic into the perspective of other outbreaks of the Modern Pandemic, the 50,000 deaths in Manchuria occurred over only a few months, while in South America 30,000 deaths were attributed to plague during the period between 1899 -1950. Although outbreaks in densely populated India have produced large death totals, mortality rates due to plague have remained very low. Despite the uniqueness of the Manchurian pneumonic epidemic, it provides evidence that human behavior can have a dramatic effect on plague mortalities.

In *Black Death Transformed*, Cohn argues that the high mortalities of plague epidemics of 1347-8 were not caused by human behavior or culture but were simply the result of the introduction of a new disease to which Europeans had no natural immunity. He argues that in the course of the next 100 years of the pandemic Europeans developed an ability to cope with the disease, which he believes indicates that they had achieved a degree of biological resistance. Since evidence to date indicates that individual humans do not develop significant, long term, resistance to plague, Cohn then argues that the Black Death was not the result of *Y. pestis*. The lessening of the large-scale impact of plague outbreaks, which Cohn considers a sign of an effective immunological response, could instead indicate that Europeans had developed cultural or behavioral adaptations that made plague epidemics more bearable and less deadly. Also, it is possible that rats were developing some resistance that resulted in fluctuations in the intensity and frequency of human plague episodes. With their short life spans, rat susceptibility to a plague can vary quickly and significantly. When fewer rats die, fewer humans would be exposed to *Y. pestis*.

Graham Twigg and J. F. D. Shrewsbury both argue that *Y. pestis* is not capable of producing epidemics with the high mortality rates recorded during medieval and early modern plague outbreaks. Twigg argues that a likely culprit for these epidemics was anthrax. Shrewsbury, on the other hand, contends that although plague was present, it was only a minor component of the epidemics that devastated England for centuries. Shrewsbury bases much of his argument on his belief that most of England beyond London was too thinly populated to support plague epidemics. In his view, only epidemics in the densely populated regions of England during the late summer months, which he considers plague months, can be attributed to the plague. Shrewsbury thus overlooks the fact that *Y. pestis* is dependent on the population density of rats and rat fleas, not on the density of humans. Shrewsbury also contends that people of the medieval and early modern periods referred to all disease outbreaks by the same terms, without differentiating between diseases.

In countering Shrewsbury's primary argument, Ole Benedictow points out that in rural areas where the proportion of rats to humans was higher than in urban areas, human epidemics would be expected to be more severe than in urban areas, an observation that is borne out in the Modern Pandemic. Cohn points out that during the Modern Pandemic in India, a plague has shown a distinct and limited seasonality. Epidemics during the Second Pandemic in England and other countries of Northern Europe predominantly occurred in late summer and fall but were not limited to these months. Although it is now generally accepted that at least some epidemics in the offseason were not primarily the result of plague, it is worth noting that in the cool climate of England plague outbreaks might have been driven not only by the late summer period of prime flea activity but also by the fact that rats, which are sensitive to cold, would have sought shelter within the interiors of buildings in the winter. Humans also likely spent more time indoors in winter than in summer, almost certainly putting them near rats and making transmission from rat to human, via fleas, possible. One key question: how well did fleas survive the cold winter months on their mammalian hosts?

Another explanation that has been proffered for the high morbidity of earlier pandemics is that the high density of human fleas and lice in the past allowed the plague to be spread from person to person via parasites. Jean Biraben and Paul Slack contend that the participation of human ectoparasites in the spread of plague distinguished earlier pandemics from the Modern Pandemic. However, Slack also notes that the plague's erratic dispersal is very different from the more universal dispersal of typhus, which, suggests that even in the Second Pandemic, rats determined the path of infection. Also, researchers are currently investigating the possibility that lice were instrumental in spreading the plague. In Modern Pandemic, there are only three known cases of the plague being spread from human to human via a parasite. It is highly unlikely that plague, spread from person to person via the agency of a parasite, contributed to the high morbidity of the Second Pandemic. It is possible that the layers of infrequently washed clothing commonly worn in the medieval and early modern period, along with bedding that was used until rotten, may have provided fleas an ideal environment very close to both their human and rat hosts. Human fleas are nest fleas and typically live in bedding and only spend time on humans while eating. Rat fleas on the other hand are fur fleas and typically live and eat on their hosts.

Twigg makes his argument against *Y. pestis* and for anthrax as the agent of plague based on several features of the Second Pandemic. As noted previously, the speed at which the disease traveled differs between the two pandemics. Twigg also argues that at the time of the initial outbreak in 1348 the rat population could not have been either large enough or widely distributed enough to have supported a plague epidemic. Shrewsbury makes the more common argument that the Black Rat, *Rattus rattus*, colonized England sometime after Bede's lifetime and well before plague arrived in England in 1348. He bases his argument on the development, and use of the terms such as large mice (*mures majores*) and rats in addition to mice, which indicate a familiarity with rats. During the earlier medieval period, mice were the targets of complaints about small animals destroying books, eating communion wafers, or eating grain, while in later medieval and Early Modern period both rats and mice were condemned as the culprits. Also, Shrewsbury points out that over time rats came to have a place in literature, page decorations, and art. He supports his idea of the growing significance of rats with an illustration of two rats hanging a cat.

Rattus rattus, which probably originated somewhere in tropical Asia, was supplanted by the Brown Rat, *Rattus norvegicus* when it was introduced into England no earlier than1728. The Black Rat is substantially smaller and less hardy than the Brown Rat and thus more dependant on humans and more easily confused with mice. It is generally assumed that Black Rats must have been present in large numbers in Europe in 1348 or there would have been no plague, and, further, that Black Rats were brought to Europe by caravans returning from the Crusades. Exactly when this rat was introduced is a matter of some debate, however.

Both Benedictow and Twigg refer to archaeological excavations in Britain in which the remains of rats were uncovered. A few rats have been found in excavations of Roman-era sites. Benedictow assumes this archaeological evidence indicates that rats were abundant; however, Twigg does not accept this assumption, arguing instead that the climate of Europe was simply too harsh to have allowed a significant population of Black Rats. He cites evidence from a study of medieval fauna taken from owl pellets recovered from the thirteenth-century level of a Roman ruin at Caerleon. The faunal assemblage examined included a wide variety of animal bones "including the house mouse ... there were no rats." If Black Rats were present in England, however, they almost certainly lived in very close contact with their human hosts because they preferred a warmer environment than England could have provided. The written record is difficult to interpret because people seem to have made little distinction between rats and mice. To further complicate the problem, the nomenclature used for animals and the distinctions made between them by medieval Europeans differ from those made today. Kellwaye, mentions that weasels along with dogs, cats, and pigs could carry plague with them as they moved place to place. In *Etymologies*, book 12, 3: 3, Isidore of Seville (7th century CE): "The weasel (*mustella*) is called the "long mouse" from its length (*telum*). Weasels attack snakes and mice. There are two kinds of weasel: a large one that lives in the forest, and another that lives in the houses of men." The descriptions suggest that it just possible that the weasels Kellwaye referred to were rats, rather than the mustelid now referred to as a weasel.

Twigg also notes that medieval commentators often associated human plague outbreaks with disease outbreaks among domesticated animals. Murrain, or cattle plague, is an ancient disease mentioned in Exodus. Currently, the term murrain usually refers to rinderpest, a disease primarily of cattle that can also attack most other cloven-footed animals, including sheep and pigs. Rinderpest is not considered to pose any danger to humans. It is not clear, however, that during the medieval and early modern periods the term murrain indicated a specific disease rather than any significant episode of mortality among domesticated flocks. Murrain outbreaks in the early fourteenth century were followed by a second series of murrain outbreaks which lasted from 1346- 1389, and these outbreaks were said to have affected not only cattle and sheep but horses, pigs, poultry and wild birds. Because murrain outbreaks occurred before 1348, it is possible to assume that these deaths were unrelated to plague outbreaks, but it is also possible to associate them with the plague or to attribute plague to the same agent or a mutation of it. Kellwaye mentions that one of the signs of an imminent plague outbreak is that "the beasts of the field, we may perceive it, (especially sheep) which will goe mourning with their heades hanging downe towarde the ground, and dyvers of them dying without any manyfest cause knowne unto us." Twigg points out that many stories link plague outbreaks to weavers and shipments of wool and fleece, and for many researchers, this connection has been assumed to be the result of fleas shipped in the wool bundles. Like these researchers, Twigg assumes that the murrain, or the high mortality among the domestic animals, was related to the human disease outbreaks.

Twigg, however, also believes that the connection between sick animals and humans was the result of anthrax, which is primarily a disease of sheep but is also communicable to humans and other animals via spores in the sheep's wool. Because wool and fleece were central to the English economy, their shipments would have been frequent and thus would have inevitably coincided with the outbreak of many plague epidemics. Although co-occurrence does not prove causation, Twigg concludes that anthrax, which is more contagious than the plague, is readily spread among cattle and sheep and is not dependent on rats, was much more likely than *Y. pestis* to have been responsible for the epidemics of the medieval and early modern periods. Also, Twigg finds many of the clinical descriptions from the medieval and early modern plague thoroughly unconvincing as descriptions of *Y. pestis* infection.

In coming to his conclusion, however, Twigg does not pay sufficient attention to social and cultural factors, particularly to household and village ecologies, nor does he sufficiently consider cultural changes that occurred between the fourteenth and twentieth centuries. In assessing the differences and similarities between the epidemics of the Second and Modern Pandemic, Twigg overemphasizes what he considers universal biological issues and underestimates cultural changes that cause people to live in different ecological relationships with their environment. People respond to and interpret similar events differently now than in the past, and also to explain events quite differently now than in the fifteenth or seventeenth century. Now that we know that rats are central to the spread of plague, it seems impossible that the English would not have noticed and remarked on the deaths of large numbers of rats had they seen them, especially in an urban setting like London. The primary explanation we have for this medieval oversight was that at that time the connection between rats and plague was unknown and thus rats seem to have gone unnoticed. Twigg cites evidence gathered in Egypt during Modern Pandemic, that shows that rats could die within the walls and never be seen. A few dead rats were found on the floor of the Egyptian house, but when the cesspit shaft was opened 53 additional dead rats were discovered. Also, because the house stank of decomposing rats and remained infective to test guinea pigs which were loosed in it, the researchers assumed that many more dead rats would have been discovered had the house been demolished. In a city like London, with many levels of construction, including considerable use of rubble construction and infill, and where hostelry walls were so flimsy that a thief could commit burglary by putting his hand through the walls, it seems reasonable to presume that during the Second Pandemic rats could have died unnoticed within the walls of buildings.

There are other reasons that the deaths of small, unimportant rats could have gone unobserved, whereas the deaths of larger domesticated animals on which the economy depended could hardly have gone unnoticed. Many of the London Plague orders, as well as medical treatises issued as early as 1592, emphasize the need for the streets to be kept clean and it is just possible that due to the extra efforts to keep the streets clean during plague epidemics, the bodies of rats were swept away and thus not remarked upon. Instead of noting the death of small animals, the English seem to have looked to their larger domestic animals for possible connections to epidemics. London plague orders demanded the control of several species of domesticated animals, including swine and cats, tame pigeons, and doves as well as various vermin, but the primary focus of the plague orders regarding animal control was dogs.

Deaths of domesticated animals might have been linked by contemporary observers to human deaths because they actually were causally linked or because coincidently, or due to environmental reasons, epidemics and murrain outbreaks of unrelated diseases occurred simultaneously. High mortalities among domestic animals could also have become linked to human epidemics because of a tendency among people to see signs, symbols, and omens in the world around them or to attempt to logically link events. Many of the signs warning of forthcoming plagues, such as those mentioned by Kellwaye and other chroniclers, seem to have been seen as omens or portents rather than logically or causally related events. Jocelin's diary indicates that he saw plague, as well as other disasters, as messages from God indicating displeasure with human behavior. This practice of linking unrelated events, especially bad or extraordinary ones, makes it very difficult to discern from medieval writings the actual interrelationships that might provide clues to the biological nature of the plague.

Nonetheless, many stories link the arrival of plague outbreaks to shipments or the arrival of wool and fleece, which had immediately preceded the epidemic. Plague and anthrax, like most other diseases, require a period of incubation, however. If wool shipments were responsible, plague flare-ups would not have occurred for at least a few weeks after shipments of fleece or wool had arrived. Thus, the stories that link plague to shipments of wool serve more to demonstrate that medieval English was trying to draw logical inferences about cause and effect in their attempts to explain plague outbreaks. Concern about the threats posed by wool shipments can be seen as extending to concern about the dangers posed by the kind of people who traveled around the country doing odd jobs or delivering goods. However, the greatest concern was expressed for people who were unknown, or who had no fixed addresses and positions. This fear of outsiders or vagabonds is clearly visible in the Elizabethan and Jacobean plague orders concerning rogues and vagabonds, which were reprinted as components of plague orders in 1636. In addition, they include regulations that limited the freedom of vagabonds and beggars, cared for the poor and assisted sailors and soldiers.

The English plague orders take very little notice of animals apart from a few references to eliminating dogs and other nuisance animals. The orders specify dogs that bark incessantly, or run loose, within London should be killed. They also mandate the control, or death of other animals including, pigs, cats, doves, and rats; however, it is clear that dogs were the primary concern of the issuers. Nonetheless, chroniclers also mentioned other animals. Boccaccio's description of the arrival of plague in Florence in 1347 that provides what purports to be his eyewitness account, mentions pigs dying instantaneously after rooting around in garbage associated with a plague victim. He also notes that anyone or animal "were it Dogge, Cat or any other" that associates with the clothes or belongings of those who have died of the plague, will become contaminated. Kellwaye mentions many animals being associated with the plague. Jenner points out that there are records of many environmental abnormalities that were said to have preceded outbreaks of plague. He cites Thomas Lodge (1558-1625) as saying that "any increase of such creatures as are engendered of putrifaction, as wormes ... flies, gnattes, eales, serpants, toades, frogs and such like were warnings of a plague epidemic. Kellwaye's book mentions other signs including a "great store of little frogs, red toades and mise on the earth abounding extraordinarily: or when in sommer we see great store of toades creeping on the earth having long tayles." Jenner further points out that these abnormalities were considered omens of a forthcoming event rather than being integrally related by cause and effect. Thus, contemporaneous explanations and descriptions of the behavior of plague epidemics are difficult to integrate with contemporary understanding of *Y. pestis* etiology that emphasizes rats and their fleas.

Although Europeans connected various animals to plague outbreaks, they made virtually no mention of rats in connection to plague epidemics. The absence of comments that connect rats or their deaths to plague outbreaks throughout the entire duration of the pandemic is perplexing; neither large numbers of dead rats nor rats exhibiting uncharacteristic behavior, receive more than a passing note. Mark Jenner concluded that this is because people of the medieval and early modern periods did not have the knowledge base to pay any special attention to rats. However, Girolamo Fracastoro (c.1478-1553) commented that one of the odd things about syphilis was its ability to spread rapidly while affecting only humans. This comment suggests the possibility that in Early Modern Europe the death of small unimportant animals, such as rats, might have gone unremarked during a plague epidemic because animals dying in conjunction with humans was considered normal. During the Tudor and Stuart periods, rat-catchers are occasionally mentioned in the literature, these people were most commonly charged with killing all manner of nuisance animals including cats and dogs, not simply rats, and were also referred to as dog catchers. These exterminators were paid a bounty for killing each of several animals that were seen as potentially damaging, dangerous, or annoying.

During the plague epidemics of early modern Europe, dogs seem to have borne the brunt of animal control zeal. From the standpoint of modern epidemiology, this attitude toward them seems misplaced because dogs are quite resistant to plague.

Although cats are fairly susceptible to plague and often come into contact with rats and their fleas both cats and dogs can spread the plague to humans.

To explain the aberration in the focus of early modern European plague fears, Jenner suggests that there were multiple reasons that dogs were the primary animal slaughtered to prevention plague, though these were only tangentially linked to any direct fear of dogs' special ability to spread the plague. Dogs' catchers were charged with killing noisy, annoying dogs, especially those that were found roaming the streets. It is possible that loose and hungry dogs were associated with urban filth because they ate garbage in the streets. In early modern Europe, rats do not seem to have been seen as dangerous as either cats or dogs, or as closely linked to plague.

Odd behavior by rats, however, was noted as a precursor to plague outbreaks in the Vedic texts of India. Benedictow argues that one of the reasons morbidity rates in India are so much lower than morbidity levels in the earlier epidemics of Europe is that Indians recognized dead and dying rats as a precursor to plague and they fled infested locations whereas medieval and early modern Europeans did not. Ibn Sina, Avicenna (980-1037), wrote that rats (or a small burrowing mammal) walked around above ground and behaved as if drunk in conjunction with plague epidemics. Thomas Lodge (c. 1557 – 1625) wrote that when "Rats, Moules, and other creatures (accustomed to living underground) forsake their holes it is a token of corruption in the same." However, his words so closely mirror those of earlier authors that it is unlikely that he is reporting events he saw, rather than merely repeating phrases from the earlier works. The occasional but repeated use of these phrases suggests that rats were seen as an omen of to disease, if not a causative factor, so one would expect that if rats behaving oddly were a generally accepted sign of plague, rat behavior or rat deaths would be more widely reported.

No one now seriously accepts Gabriele de Mussis's story, which explained that the plague was introduced into Messina by soldiers who fled fighting in the Crimea. It is discounted not only because it has been proven that de Mussis was not an eyewitness to the events, but also because it is hard to imagine how any disease could have remained rampant, and infective, without killing sailors who harbored it during a voyage of several months. Also, this story makes no allowance for incubation periods once the disease was introduced in Messina. The stories linking the wool trade to plague have been less seriously examined, yet they are similarly flawed.

For example, the story that explains how the town of Eyam became infected is still often accepted as reasonable. According to the story, the plague erupted in Eyam immediately after the village tailor, George Vicars, opened a shipment of cloth sent from London. Although it is possible that a disease could have arrived with the cloth, it is not possible that the tailor opened the box and immediately became ill with the plague. Philip Race indicates that he finds a permutation of G. R. Batho's version of the story, in which the tailor was infected by the cloth shipment but the onset of the disease was slower, and thus more in line with a modern understanding of disease incubation, a reasonable one. It is far more likely and simpler, however, that plague entered the Eyam rat population unnoticed at some earlier date and coincidently became epidemic when the cloth arrived. These stories of how epidemics began may accurately present the relative time of events, but they do not demonstrate cause and effect. Twigg tends to accept plague stories like these that link trade in wool to plague outbreaks because these such fit his theory that the epidemics were caused by anthrax. Researchers who favor the theory that plague was caused by *Y. pestis* also find the connection between wool trade and plague reassuring, but they explain the relationship by postulating that infected fleas were shipped in bales of wool. Slack finds that "there is ample contemporary reference to particular individuals or bundles of the merchandise being responsible for initiating a local outbreak." I find the evidence less convincing. While contemporary scientific research has not eliminated the possibility that the plague was spread in bales of cotton or wool, it has suggested that shipments of grain were more likely to have been responsible for transporting the plague. Also, contemporaneous observers who associated plague with shipments of cloth did not know about disease incubation periods and therefore did not make allowances for it in their cause and effect calculations. They would likely have connected unrelated events to plague outbreaks.

Scott and Duncan, like Twigg and Shrewsbury, find it difficult to believe that *Y. pestis* could have produced the high mortalities recorded in many of the epidemics attributed to plague, but unlike Shrewsbury, they do find the mortalities reported during the Second Pandemic credible. Although Scott and Duncan esteem the work of Twigg, especially his critiques of the idea that the epidemics were produced by *Y. pestis*, they do not find Twigg's solution convincing. Instead they have postulated that the medieval and early modern epidemics referred to as pestilence were the result of a variant of hemorrhagic fever, not ebola but something similar to it, presumably something that no longer exists.

Conclusions

This examination of the literature addressing plague pandemics shows that though plague outbreaks have been studied and written about for centuries, many questions remain. Contemporary concerns about epidemics were not succinctly or addressed by contemporaneous observers and plague survivors. It is equally clear that contemporary historians and microbiologists have not reached a consensus about how to interpret either the information provided by historical treatises or the information provided by contemporary scientific inquiry. DNA analysis has not yet provided conclusive answers. Equally importantly, few contemporary historians have examined both contemporary scientific information and the historical conditions under which plague thrived and spread readily. These conditions include the presence of enormous quantities of garbage and other refuse in and around houses and in streets, construction methods that provided rats habitation within walls and attics.

The Medical Context

The plague that swept across Europe in 1348 had a definite influence on two European institutions: The Church and the medical establishment. As the most powerful institution in medieval Europe, the Church did not fare well in the wake of the plague onslaught because the plague was perceived first and foremost to be a result of a failure of faith. It was the only institution with the scope to battle the universal threat plague presented, but due to the nature of the threat, the Church inevitably failed. People believed that the Church should have protected them, or at the very least it should have warned them of impending doom, but it provided neither warning nor protection.

Medicine and the Church were by no means in opposition, but when the religious community failed to mitigate or control the spread of plague, physicians and surgeons stepped into this vacuum. The failure of the religious community provided physicians and surgeons an opportunity to increase their relative political prestige. The Church attempted to take an active role; through its bishops, the Church called for community prayers and processions, but civic leaders turned to the medical community for assistance in combating the pandemic.

Physicians and surgeons acted individually and in groups, on their own as well as at the request of rulers, and they wrote treatises that provided suggestions for the appropriate responses to the massive catastrophe that was ravaging Europe. These early treatises indicate that pestilence was seen as an attack on the entire social fabric. As Guy de Chauliac put it, "charity was dead and hope crushed." English descriptions of the plague's arrival emphasize problems such as those created in the economy and high mortalities that left no one to care for the sick or bury the dead. Although the medical community did not succeed in preventing plague outbreaks, the concept of public health developed in the wake of the plague's destruction, and eventually it became a central aspect of English plague regulations. During the Second Pandemic, the medical community strove to understand and control the plague using both pieces of knowledge gained from texts and personal experience, though they met with little success. The medical treatises written during the first few years of the pandemic were both a reflection of new experiences and the product of the medical tradition.

One of the earliest plague treatises, written by Gentile da Foligno who died of the pestilence in June 1348, directly addresses the necessity for a public response to the plague by recommending that the civil authority appoint a committee to meet with physicians to "manage public affairs in so far as the safety of the people was involved." Foligno's treatise was only one of several early tracts addressed to civil authorities. In the battle to combat the plague, the theme of public service appears repeatedly. Because the medical establishment did no better than the religious establishment at preventing the spread of plague or at treating patients, members of the medical community quickly came to be criticized and ridiculed for their failings and greed. Still, medicine as a profession was not repudiated.

In 1348 when plague entered Europe, the medieval medical tradition was well established across most of the continent. The works of Rhazes (864-

930), and Avicenna (980-1037), and Galen of Pergamon (ca 130-ca 200AD), as well as some portions of the Hippocratic corpus, were widely distributed and known. Medieval European physicians had no experience with widespread epidemic mortalities, but the medical literature that physicians relied on did include the works of ancient physicians who had witnessed epidemics. Galen, the most revered of the ancient medical writers, had survived the disastrous epidemic at Aquileia in 168, which may have killed one-third of the population, and Razhes had described and treated smallpox and measles. Thus, medieval physicians did have a limited conceptual background of widespread devastating diseases, however, their personal experiences suggested that this plague was entirely new and uniquely devastating. Nonetheless, medieval physicians attempted to understand the plague in terms of the fevers commonly addressed in the canonical medical texts.

Medieval medicine was based on a Greco-Arabic tradition and most medical texts were based on a few translations of Greek medical texts. From the fall of

Rome in the fifth century until Constantinus Africanus (d. 1087) translated Greek medical texts from Arabic into Latin in the eleventh century, Europeans had few medical works from the classical past. The works that Constantinus translated became the core of the medieval medical canon, which was gradually enlarged as more works were translated from Arabic into Latin and as new texts were produced using commentaries written on the older works. Throughout the Second Pandemic, the essential medical canon increased substantially. It expanded as more extensive portions of works by Galen, Aristotle, and others from the Hippocratic tradition were translated into Latin, new versions of old texts were found, texts were printed in Greek and new, more complete, elegant, and accurate translations of Greek medical manuscripts were produced. The focus of most of this work was the rediscovery of lost medical knowledge rather than the addition of new empirical knowledge. Nonetheless, along with the new translations, new botanical works with more naturalistic drawings were printed. In some cases, this may have meant that botanical compounds, or Galenicals, were being formulated more accurately as people came to re-identify various plants. Although the goal was to rediscover ancient knowledge, the new texts ultimately encouraged challenges to the medical canon because they presented a broader more diverse vision of Greek medical concepts. Although the new translations were better and much more complete than the early ones, some of the early translations remained in use well into the sixteenth century. Thus, conflicting ideas about medicine and plague coexisted for centuries.

Despite improvements to the medical canon, the core medical philosophy remained essentially unchanged in the centuries between Constantinus' work at Salerno and the time plague arrived in Europe. According to the humoral medical paradigm, illness, or a lack of health, was caused when the balance of the four humors, black bile (Melancholy), yellow bile (Choleric), blood (Sanguine), and Phlegm (Phlegmatic), were disturbed. Each of the four humors is associated with one of the four elements: earth, fire, air, and water. In turn, individual foods and drugs were each associated with an element and its humor. The disease was also thought to be the result of miasmas, or corrupt air, which could invade the body via breathing and open pores, and destroy its natural balance. The onslaught of the plague initially disrupted this understanding of disease because of the plague's rapid spread, which was immediately seen to be contagious. Throughout the pandemic, however, because the plague's means of dispersal remained inexplicable, ideas of contagion merged with those of miasma and a general humoral understanding of the body. Thus, medical treatments for plague tended to follow the logic of humoral medical understanding, and the primary emphasis of pharmacology remained to bring the body's four humors into balance.

Theoretically, a humoral understanding of disease suggests that inter-personal differences in disease symptoms are the result of variations in individual humors.

However, by 1348, Rhazes's had distinguished between smallpox and measles and noted that smallpox was most common in the young. Razhes's work makes it clear that even before the plague entered Europe, some individual diseases were recognized and distinguished from one another even within a humoral understanding of medicine. After the plague entered Europe, Europeans understood the plague to be a unique and terrible disease, which also suggests that they could conceptualize diseases as distinct entities. Although symptoms were not a primary factor in understanding or describing diseases because they were expected to vary with the complexion of the patient, shortly after the second epidemic Guy de Chauliac described specific symptoms typical of the disease.

When the plague first appeared in Europe, many attempts were made to understand how it spread and specifically what part of the body is most affected. Shortly before the first wave of the plague pandemic, human dissections had begun to be a feature of some universities' medical curricula, particularly in southern Europe. After the plague appeared, requirements for dissections became more common; universities including Bologna wrote course plans that included annual dissections as part of the curriculum. In the wake of the first plague outbreak at Avignon, the Pope requested the dissection of some plague victims to discover the nature of the disease. The results demonstrated that those who died quickly spat blood and had an infected lung. It was only after the plague entered Europe that "public officials other than educational authorities" ordered autopsies. Plague probably was a factor in the increase in dissections, which reflects an interest in understanding the physical body better to address epidemics such as the plague. Nonetheless, this new anatomical knowledge produced very little information to change the understanding or treatment of plague, despite the increase in dissections during the fifteenth and sixteenth centuries, and increased accuracy and detail of anatomical drawings.

During the initial outbreak between 1348 and 1350, physicians wrote a considerable number of tracts addressing the problem of plague. Their treatises demonstrate a shared medical philosophy, yet they also record the beginnings of a debate about the mechanisms by which plague spread. In general, the early treatises addressed three basic issues: the underlying root causes for the plague, how to avoid the plague, and how to treat patients who have the plague. Though many plague texts were produced, none of them became central to medical education, which continued to center on traditional texts and concepts. The most extensive section of the early treatises was devoted to a consideration of the causes of the epidemic and the least attention was devoted to treating plague victims because, as several of the authors note, "the disease almost always has a fatal termination." In later plague treatises, consideration of the causes of plague became a less significant portion of the whole.

Most of the early medical treatises were composed in southern Europe, where the largest medical schools were located, and none of the surviving treatises written in the first few years of the pandemic was written in England. Several of the later English plague treatises were translations of earlier continental documents. For example, the fifteenth century *A Little Book for the pestilence* was an English version of Bengt Knutsson's version of Jean Jacmé's 1364 treatise. Some of even the earliest tracts were written in the vernacular. This practice became increasingly common after the 1450 invention of the printing press, which made it possible to address the lay public's desire for health and medical information. The production of a huge number of medical texts targeted to the learned layperson as well as to medical professionals suggests that the medical community saw the problem of treating large numbers of people with the plague was too large for university-trained physicians to handle on their own. The treatises produced by the expanded interest in medicine and medical inquiry, however, proved of little value in treating victims of the plague.

Anna Campbell divides the earliest treatises' descriptions of the causes of plague into two groups, "the general and remote and of the particular and near." The category of "general and remote" refers to reasons or causes for the disease's existence while "particular and near" refers to how the disease enters individual bodies. An example of the latter is that authors addressed ways of avoiding or preventing the corruption of the air and thus infection by the plague. Remote causes included some sort of astrological occurrence that rendered the air putrid and produced pestilence. The track written by the faculty of the University of Paris was the most precise and detailed in its description of exactly how astrological events influenced the spread of the plague. "On 20 March 1345 at one o'clock in the afternoon, occurred important conjunction of three higher planets in the sign of Aquarius, which with other conjunctions and eclipses is the cause of the pernicious corruption of the surrounding air, as well as a sign of mortality, famine and other causes not connected with the present subject." The Paris Medical Faculty's explanation was the basis for many tracts produced during the next hundred years. An Italian version described the events as a solar battle that occurred over India and the Great Sea, which produced toxic mists in the atmosphere by sucking up corrupted water, and then explained, more specifically, the actual point of origin and spread of the plague. Almost universally, the near causes of plague were understood to be a corruption of the atmosphere; however, there was less consensus regarding the description of corrupted or putrid air.

The miasma of putrification of the air that was thought to cause plague was often described as a vapor. The authors of the early medical tracts said that this putrification could be carried long distances by lightning and winds and they considered winds from the south to be especially dangerous. They also noted that the vapors could escape from enclosed spaces such as ships' holds or storehouses, wells, lakes or ponds, or arise from unburied corpses or other foul-smelling material. Thus, fires came to be seen as a means to counteract the vapor by creating smoke, especially by burning sweet-smelling herbs, which could counter harmful corruption of the air. Additionally, fires were thought capable of materially altering harmful vapors and preventing the spread of the plague. Fire and smoke remained a recommended weapon against the plague throughout the entire pandemic. In 1348, Gentile said that fires in the street would combat plague by destroying urban smells and Pope Clement VI was protected by large fires, and as late as 1665 during the Great Plague of London street fires were built in London streets to combat the pestilential corruptions of the air. During this last epidemic London street fires met with some resistance because, in addition to being very expensive, critics did not think that the fires could produce enough heat to materially change the bad atmosphere. Nonetheless, the use of fire and smoke to combat plague, and even the argument against the fires, suggests that throughout the Second Pandemic, battling atmospheric corruption was an accepted method of fighting the plague.

Many early treatises emphasized the means of avoiding plague both because physicians and surgeons had no clear idea of how to treat plague and because preventing rather than treating disease was central to the humoral understanding of health. Over time, plague treatises came to include more recipes for medicaments; however, many of the recipes also were recommended for plague prevention. Much of the advice as well as the recipes were repeatedly reissued under new authors. Aside from providing explanations of the causes of plague, the first plague treatises relied on the wisdom from revered authors of the past. Authors admitted that they were dealing with what appeared to be a new phenomenon, if only because it was of a wider scope than any previous epidemic; nonetheless, they attempted to apply old remedies to the new situation. Thus, little of their medical advice seems to be directed specifically at preventing plague. Although the advice was refined and modified, moderation and behaviors developed to combat earlier diseases now presumed to be malaria, continued to be recommended against the plague throughout the Second Pandemic.

In the sixteenth century, Simon Kellwaye's treatise illustrates this conservatism. His treatise includes advice to be moderate in eating, drinking, and exercise, along with advice on what environments are likely to be the healthiest. Kellwaye says that there are three primary methods of avoiding the plague. The first is prayer, the second flight and:

> *The third means consisteth chiefly in three poyntes which are these: Order, Diet, and Physicall helps. For the first you must have a care that your house bee kept cleane and sweete, not suffering any foule + filthy clothes or stinking things to remaine in ... and in sommer season to deck your windows and strawe your floores with sweete and*

> *holsome herbes, floures and leaves as Mintes, Balme, Pennyriall ... and such like, for your windowes, your floores to be strawed with greene Rushes and Mynts, Oken and willow leaves, Pine leaves and such like: your windowes which stand towarde the North and Easte, doe you alwaies keepe open in thedaye time, (if the aire be cleare and that no infected and that no infected and unsavory smell be neere the same) as fogs, doonghils and such like, and every morning before you open either your doors or windowes as also in the evening when you go to bed cause a good fire to be made in your chamber and burn some odoriferous sweet perfume.*

Medicines recommended preventing plague were formulated primarily in an attempt to bring the four humors into balance and thus to make people less susceptible. Many of the same medicines were also recommended for patients already infected by plague, although the plague tracts also emphasized that as soon as plague symptoms became visible, other treatments were needed to eliminate toxic humor.

Various forms of purgatives, bleeding, sweating, and emetics were recommended to balance the humor and to rid the body of toxins. Also, surgical and medical procedures such as poultices were recommended to remove toxins and drain buboes. Recipes for the ingested medications or nostrums could be quite complex, and some such as theriac or treacle, included as many as 80 ingredients. The sheer volume of ingredients included in many treacles makes the simplicity of the four ingredients of "Diatessaroum triacle," which was recommended as a "Medicament expulsive" in the medical advice printed along with several of the plague orders, stand out. The ingredient lists included with the plague orders are also considerably less extensive than those included in many of Kellwaye's recipes, which suggests that the authors were doing their best to provide, as required, "sundry good rules and easie medicines without charge to the meaner sort of people." Despite the diversity of ingredients, a few occur repeatedly. Ingredients favored as plague preventatives include rue, figs, walnuts, butter, and any number of mostly sweet-smelling herbs. Preventative measures included both those that were worn and those that were taken internally; many odoriferous ingredients were used to make various forms of nosegays, which were sniffed to ward off plague and other dangers produced by bad smells. Many of the same ingredients were also burnt to fumigate houses.

The logic behind odor prevention measures is quite simple. It was believed that bad smells carried the danger of disease and illness; that, in effect, the bad smells produced by rot and decay created the miasma, which produced and diffused disease. Thus, smelling bad smells meant that you were being exposed to the miasma and infection. Given that people had recognized that disease often broke out in the aftermath of battles, amidst the stench of decaying corpses and rotting garbage, this was not an illogical assumption. Thus, using sweet-smelling herbs to purify a room or to carry when walking in a potential plague-infected area, were recommended preventative measures. Kellwaye's plague treatise of 1593 provides a list of herbs for clearing a room that includes mint, pennyroyal, lavender, thyme, red roses, as well as numerous recipes, each of which includes a list of several alternative ingredients, for making various sweet-smelling concoctions to be worn near the heart. Carrying nosegays continued to be recommended during the last major plague epidemic in London. Even though he does not seem to put a great deal of faith in the value of these scents to provide protection, Pepys mentions being forced to buy tobacco to chew and sniff when his business forced him to walk near a house marked by a white cross as a plague-infested house. Also, after tobacco was introduced to England, smoking tobacco came to be recommended as a plague preventative. Indeed, as the plague came to be associated more closely with the poor, its association with bad smells increased rather than decreased. It was linked not only with the smell of rot and decay but also with fetid air common in overcrowded housing conditions.

Treatises such as Kellwaye's provided information on how to draw poisons out of the body as well as on diets to prevent plague, based on a traditional medical understanding. Kellwaye emphasizes moderation but also considers the relative heat and moisture values of various foods. His suggestions favor food that is drying, but he also warns against meat, such as goats and water birds, because they are hard to digest. He also suggests avoiding lamb because it is moist and he brings up an interesting argument on the merits of eating fish. Kellwaye notes:

> *there are some authors which holde opynion that fish is more better to be eaten then flesh in the great fervent heate of the yeére, because they doe make a mold colde bloud int the body than flesh: another reason is because they doe live under water, they are not infected with any contagion of the ayre, as beaste and Byrdes may be and therefore more wholesome but in my iudgement flesh is more wholesome because it doth bréde a more pure*

and fine Juyce in the body, than any fish whatsoever.

Although the humoral system remained the dominant explanatory system for human health and disease, treatises written in the vernacular for a non-medical audience reflect a simplification of the humoral system, which was too complicated for non-medical people to master and interpret. Medical advice on plague issued in England, largely for non-medical people, during the last hundred years of the Second Pandemic makes very little reference to the differing medical needs of people with different constitutions or complexions, or to seasonal variations. For example, although Kellwaye uses a variety of humoral terms, he often divides people into two general categories, employing common terms such as "strong and rusticall" or "daintie, and idle bodies." Medical advice printed in conjunction with the plague orders also includes a few recipes with alternative ingredients for summer and fall as well as recipes targeted to the poor. Interestingly, Kellwaye's advice on smallpox includes a more detailed discussion of how humors affect the course of that disease than his advice on plague contains.

Many of Kellwaye's recipes provide dozens of alternative ingredients, and he does not indicate whether any specific combination of ingredients is preferable. Many of the recipes recommended by the College of Physicians, which were printed along with the plague orders, are similar to those included in Kellwaye's book, although they are simpler. The dissimilarities between their recipes are linked to the goal of the medical advice, printed along with plague orders to provide "sundry good rules and easie medicines, without charge to the meaner sort of people." Despite the simplifications, an examination of recommended ingredients listed in these publications indicates that no single ingredient or method of treatment dominates.

Throughout the latter part of the Second Pandemic, treatises recommend avoiding plague-infected people and places, but the rationale that linked places of foul smells, and fog or dampness, with contagion, was still largely based on a humoral and miasmic understanding of the disease. Kellwaye notes that plague can be found in ships' holds as well as in crowded jails and rooms in the back alleyways of cities. The implication is that it is the smells created in these locations that produce plague rather than that plague simply moves from one person to another. In chapter 12, "Doth shewe what you must doe when you goe to visit the sicke," Kellwaye explains what a surgeon or barber who is going to visit the room of a patient ill with plague should do. First, he recommends that the surgeon should make sure that a large fire is lit in the room with the patient and that some herbs are burnt. Kellwaye also recommends that the medical man should hold a spicy lozenge in his mouth and have an additional sweet-smelling concoction to sniff, and he should not stand between the patient and the fire, because the fire will draw the toxic vapors. Also, Kellwaye recommends that the surgeon should have someone else expose the patient's arm to the air before the patient is bled so that the dangerous sweat has a chance to evaporate before he approaches the patient.

The implication of Kellwaye's advice seems to be that, whatever the nature of the infectious agent, it is comprised of smell and moisture. Kellwaye's discussion of a different disease, smallpox, also demonstrates that the early modern European understanding of disease dispersal and terminology differs from a modern epidemiological understanding. Kellwaye enumerates four levels of causation for smallpox and measles and calls them both diseases "hereditable" and "infectious." He describes their immediate cause: "the conjunct cause is the menstrall bloud which from the beginning in our Mothers wombes wee receaved, the which mixing itself with the rest of our blood, doth cause an ebulition of the whole." He explains that differences between smallpox and measles are the result of differences in heat and moisture in the blood but he did not see the differences as fundamentally significant because "they are both one in the cure."

Frequently European physicians selectively used the work of their predecessors in the way Kellwaye used observations on the difference between smallpox and measles. He begins the chapter, *Warnings of the Plague to Come*, by referring to Avicenna, or ibn Sina, and the way ibn Sina linked seasons and seasonal abnormalities to health and illness. Kellwaye's use of ibn Sina illustrates one of the major problems faced when examining plague texts, namely that it is difficult to distinguish between the details the writer is reporting from personal experience and those that the writer copied from earlier materials. Kellwaye does not distinguish between the signs warning of an imminent plague outbreak that was commonly reported in England and those that were reported in the ancient past. Typically, plague treatises incorporated large sections of earlier writings without attribution.

As we have seen, the basic precepts of humoral medicine as defined by Galen dominated medical explanations of plague, which emphasized the variability of symptoms between people as the result of differing humoral balances. Plague, however, came to be seen as being caused by poisonous material around the heart. Although the diffusion of plague proved impossible for physicians to explain, they understood that it had a different cycle and intensity than the scourges with which they were familiar because it spread very rapidly across the entire continent and killed many people quickly, yet they had no alternative concepts to apply.

The severity of plague epidemics was sometimes blamed on a diet. John Lydgate (ca. 1371-1449) wrote "Do not eat meat out of greedines, and abstain from eating fruit." Writing about a century later, Kellwaye also recommended eating in moderation. Plague was often linked to eating fruit and to extremely bountiful harvests of fruit, which meant that the poor were eating large quantities of fruit. Although eating large amounts of fruit is unlikely to have contributed to high levels of plague among the poor, it is possible that good harvests could have affected plague patterns by fostering larger than usual numbers of rats.

Throughout the 300 years that plague epidemics ravaged England, the medical profession did not develop a standard treatment protocol. Kellwaye's treatise includes many recipes, most of which include many possible variations. A recipe for "A Good Glister" calls for one handful of mallows, beets, violets, or red fennel, to be combined with one dram of seeds of fennel, anis, or coriander. The water boiled with these two ingredients is then to be mixed with another set of ingredients, again with a few choices. The choice of alternative ingredients in these recipes supports my prior contention that no one set of ingredients had become the standard preventative or treatment for plague. The Good Glister is specifically mentioned as a preventative that is good in times free of plague as well as during plague visitations.

Many plague medicines, like this one, were designed to assist the body's ability to resist disease. Attention was focused on attempting to prevent plague outbreaks or avoiding the disease altogether, rather than on curing the ill.

The discrepancy between early descriptions of plague as contagious and the long-running debate about whether plague was or was not contagious was the result of several factors. Medical theory of the Middle Ages and the early modern era did not endorse the concept of contagion, and yet plague was consistently described as being spread from person to person and from the community to community, especially during the first wave of the Second Pandemic. Despite the consistency produced by reissuing plague tracts repeatedly, various and somewhat divergent ideas were expressed during the

Second Pandemic about the essential nature of the plague and how it was spread. In part, these inconsistencies originated with differences evident in the initial plague treatises, and in part, they were the result of dissonance between theoretical precepts and the observed behavior of plague.

A miasmic theory of disease could explain the plague's extensive diffusion, but as plague spread, it quickly came to be linked to interaction with other people. Gabriel de Mussis' story traced the spread of plague west from Kaffa, and linked the transmission of plague to contact or proximity with dead victims of the plague. As a result, sailors who were accused of introducing plague were quickly banished from Italian port cities. Towns such as Venice and Milan instituted quarantine to prevent the introduction of the disease.

Though the miasma theory principally explains sickness as an alteration or pollution of the air, it does not preclude an explanation somewhat similar to a modern understanding of contagion. The corrupted air came to be seen as being carried from place to place in fabrics or animal fur so that animals and people were seen as carrying the infectious agents around.

During the Second Pandemic, the medical establishment gave de facto approval to the concept of contagion by recommending various quarantine measures that would be most efficacious against a contagious agent. However, medical tradition and the unpredictable and mysterious manner in which plague waxed and waned, and the way it moved, apparently randomly from one site to the next, prevented a rigid acceptance of a fully developed contagion-centered theory of disease spread.

However, there is considerable evidence that especially during the early in the Second Pandemic plague was considered contagious. For example, Giovanni Boccaccio (1313-1375), and Guy de Chauliac (c.1300-1368) referred to the plague as highly contagious. Although most of the earliest English chroniclers emphasized the universality of the great mortality rather than its contagiousness, Geoffrey le Baker (*fl.* 1326–1358) wrote that people believed "that the breath of those who had lived among the dying would be infectious." Boccaccio, in the introduction to *The Decameron*, describes plague as transferable "from the sick to the sound, in a very rare and miraculous manner... The quality of this contagious pestilence ... and catching it one of another, either men or women." Later writers such as Simon Kellwaye and Thomas Dekker also described as being spread from person to person. English cities attempted to prevent entrance to travelers from plague-infected regions, and English plague orders from at least the sixteenth century required infected houses to be shut up, with their ill inhabitants and anyone who had been in contact with them locked inside. Early writers appear to mean that plague spread quickly from person to person, but they did not specify how it spread, for example as a miasma trapped in clothes, or as poisonous humor escaping from the ill. The definition of contagious appears to alter so that although plague retains its description of being highly "contagious" throughout the Second Pandemic, people's understanding of what that meant did change. Simon Kellwaye begins his 1593 book with a chapter entitled "What the Plague is," noting that while there has been considerable disagreement about what causes the plague, most agree that it is a "pernicious and contagious fever." Nonetheless,

Kellwaye fears transmission via various nasty odors at least as much as being in the presence of a person sick with the plague.

Under the impact of syphilis, which struck in the last decade of the fifteenth century, the medical tradition began to change. By 1500, syphilis was generally understood to be a contagious disease spread by sexual contact. In their efforts to treat syphilis patients, physicians quickly began to expand their repertoire of drugs beyond the relatively harmless, if mostly ineffectual, simples and Galenicals to more dangerous concoctions which included heavy metals such as mercury. Despite limited evidence of mercury's efficacy, it continued to be used to treat syphilis into the twentieth century. Treatments used on patients with syphilis quickly became very aggressive in a way that the treatment of plague victims did not. Plague victims typically died quickly, while people with syphilis lived long enough to be treated vigorously. Because plague is an epidemic disease, during outbreaks there simply were not enough doctors to provide patients intensive treatments or detailed individual humoral assessments.

In contrast to syphilis treatments, the medical response to plague remained relatively constant and relatively non-invasive. The initial plague tracts recommend bloodletting only in the initial days of exposure, and both Kellwaye and the physicians' advice accompanying plague orders recommended bloodletting only if the patient was strong enough, and only during the initial day or two of the infection, and even then only if plague sores had not yet appeared.
As late as the 1720s, when Dr. Richard Mead (1630-1699) wrote *Short Discourse Concerning Pestilential Contagion* because the plague was ravaging Marseilles, there was still no consensus as to whether plague was a disease spread by a miasma of the air or by person-to-person contagion, or by some other means. For example, doctors in Marseilles declared that plague had infected the city and that plague was contagious, while at the same time Marseilles' officials concluded that plague was caused by a variety of different factors including a poor diet, rather than by a contagious disease. Similar disagreements had taken place in seventeenth-century London when the Crown

demanded that fairs and trade be curtailed because the plague was rife in the city. This disagreement in London, however, was couched not in terms of whether plague was contagious or not, but rather in terms of whether plague was caused by poisoned, or corrupted, food and goods, corrupted air, or humor, which could cling to a person and thus be moved from one place to another.

It is possible that the stories of the plague's extreme contagiousness, which developed when it first appeared in 1348, prevented a thorough reappraisal of its actual contagious potential, based on its behavior. Physicians, surgeons, and chroniclers observed that the manner of plague's spread was erratic and sporadic and was not easily linked to the person to person contact, and so concluded that plague was, as it had been originally described, especially contagious. Europeans' inability to form a consensus on the plague's etiology probably owes more to the complexity of how plague, caused by *Y. pestis,* is transmitted than to a failure of the observers and medical practitioners.

Throughout plague epidemics in Europe, doctors

and civil authorities frequently collaborated. The titles or subtitles of plague treatises often indicate that they were published for the good of the country. From the first year of the Second Pandemic until as late as the 1720s, when the plague was threatening Marseilles, plague treatises were addressed to public authorities. During London's last century of epidemics, Royal plague orders were issued in combination with medical advice and the English government asked Dr. Richard Mead to provide guidelines for controlling the plague in England.

The title page of Kellwaye's 1593 *Defensative Against the Plague* states that it was "published for the love and benefit of his country." Kellwaye's note "to the friendly reader" states that he chose to write in simple English prose rather than in Latin or elegant stylized verse so that the poor, who were most in need of medical help, could read and understand his advice. He admits that his use of such an unassuming style might make some people question his professionalism; however, Kellwaye says that getting his medical advice to those most in need was more important than his reputation. To make his recipes easier for common people to follow, Kellwaye includes a table showing conversion rates between a Latin measuring system and a less formal measuring system. Despite his concern for the common people, he included a recipe or two written in Latin.

In the early treatises, it was often assumed that a medical professional would be responsible for reading and interpreting as well as implementing medical advice. The instructions provided by Kellwaye seem to suggest that he assumed patients are cared for by around-the-clock attendants with advice from physicians. While the medical advice included in the plague orders does contain references to "the patient," the advice is simpler than Kellwaye's and the plague orders frequently make use of the word "you" and thus seem to be targeted directly at patients.

Although the advent of the plague did not produce dramatically new treatments or a new medical philosophy, it did result in an increased interest in medicine. Self-diagnosis texts had begun to become more common even before the invention of the printing press. Nonetheless, the plague greatly increased interest in and the production of medical information. These books were often printed in the vernacular, suggesting that they were of interest to a broader section of the population than elite university-educated physicians.

When the plague first entered Europe, it was immediately seen as both a social and a medical catastrophe, and the medical establishment hastily attempted to assist in the prevention of the disease. Despite little initial success, the medical community remained a partner with the government in attempts to prevent and control plague outbreaks. With its active engagement in the fight to control the plague, the medical community developed some consensus about what triggered it, how it was spread, and how to treat it, although many of the details remained in dispute. Plague appears to have been consistently described as uniquely contagious and dangerous even though, after the initial wave of the pandemic, epidemics no longer spread like wildfire across vast regions of Europe. Instead of spreading across the continent, subsequent outbreaks erupted sporadically and randomly. The medical community's inability to reach a consensus, or to come to terms with plague and the way it spread, is in itself evidence that plague was caused by *Y. pestis* or another disease with an equally complex mode of transmission.

Material Culture, Social Customs and Civic Regulations

During the centuries that plague-ravaged Western Europe, society underwent a significant metamorphosis. Transformations in culture and society were both the direct result of changes in population and material wealth and the indirect result of societal accommodations to plague. At the time plague reached England, its society was experiencing stress and pressure due to crop failures and overpopulation, but the plague's impact was extensive and far-reaching. Additional societal changes arose as a result of efforts to combat the plague. Although many cultural changes only became fully developed throughout the epidemic, some of these shifts began immediately. As a result of the initial epidemic of the Second Pandemic and the population reduction that plague produced, material possessions were divided amongst many fewer people, and, more importantly, the land was divided among fewer individuals. Thus, per capita wealth and income increased. This new wealth allowed many changes, but even before these changes were manifest, cultural behaviors changed to combat the plague.

One of the immediate, if temporary changes plague wrought was a change in burial and funeral procedures. Many chroniclers described social norms, especially burials, as having been disturbed by the plague. Early English chroniclers noted, "that the living was hardly able to bury the dead." The author of *Historia Roffensis* wrote that "men and women carried the bodies of their little ones to church on their shoulders and threw them into mass graves." Despite such comments, excavations of mass graves suggest that bodies were neatly laid in the ground, even in the mass burials. The chroniclers' comments suggest that they were distressed by the whole funerary process. Boccaccio wrote more directly that during the high mortalities, funeral behaviors underwent a considerable change. He noted that funerals became less a matter of public ceremony and display of sorrow and loss, and more a hasty matter of getting the dead into the ground. "Very few would accompany the body to the grave, and they not any of the Neighbours, although it had beene an honourable Citizene, but onely the meanest kinde of people such as were grave-makers, coffin-bearers, or the like that did these services onely for money;" they performed this duty as rapidly as possible. Burton, one of the few English writers to mention funerals, wrote that in many places chaplains survived until after they had performed the necessary funerals, and then they too died in great numbers. Unlike Boccaccio, many of the English chroniclers were monks, which may have given them a different understanding of the plague's effect on civil customs.

Throughout the plague epidemics, commentators mention the brevity and lack of ceremony invested in funerals during plague epidemics. Most of these descriptions emphasize the lack of solemnity given to funerals, issues such as the ban on funeral processions, and the creation of mass graves. Geoffrey le Baker (*d.* c. 1356) says that in the wake of plague "hardly anyone dared to have anything to do with the sick," which suggests that funeral gatherings may not have been common. However, as prints from 1665 make clear, funerals were not held completely without processions of mourners. Prohibitions on public funeral gatherings and processions are a rational response to a contagious disease, but it is unlikely to have had any effect on the spread of a vector-borne disease like the plague.

Virtually ignored by most commentators is the one aspect of the change in traditional funeral rights that might have reduced plague transmission, namely, a profound change in the more private aspects of death rituals. Before the onslaught of the plague, people congregated together in the home, or at least at the front door of the deceased, to express their sorrow. London plague orders from 1592 stipulated that no funeral ceremonies were to be held in the home of the deceased until 28 days have passed since death in the house. Indeed, London houses where people had been infected with plague were ordered to be shut up for from 28 to 40 days, with access granted only to caretakers. The first printed plague orders, which probably date from 1578, state that "if there be any doubt that the masters and owners of the houses infected will not duely observe the directions of shutting up their doors especially in the night then there shall be appointed two or three watchmen." These guards were authorized to place people who ignored the regulations into the stocks. Plague orders from 1625 specifically stipulated that "at the burial Dinner, or attendance on the Corps, or other solemnitie of any Dying of infection, there shall be no assembly of people in the house where such person shall Die within the time of xxviii after such death." Discontinuing the tradition of friends and family meeting together in the houses of plague victims could have had an appreciable impact on the spread of any infectious disease, particularly a vector-borne disease spread by household vermin, by reducing the number of people exposed to the house and its fleas.

Providing guards to enforce household quarantines did mean incurring expenses, but the changes in funeral rituals cost no money and seem to have been an almost immediate, painful, adaptation to high mortalities, whereas many other changes recommended combating plague epidemics were expensive and were only gradually implemented. Although it was almost universally accepted that the material possessions of plague victims, especially clothing and bedding, posed a threat to healthy individuals, Carlo Cipolla cites an incident in which two men were caught stealing from the empty houses of plague victims. Such thievery suggests that despite the widely accepted dangers of plague-infected material, poverty and dearth drove people to risky behavior.

In the years before the plague arrived, the population of England reached a peak that was not surpassed until late in the fifteenth or early sixteenth century. Although total population figures can only be rough estimates, the evidence that does exist suggests that as late as 1500 the population of Europe had not yet reached the population levels of 1300. By 1300 people had begun to cultivate very marginal lands, and England's peasants were approaching the maximum level of food production, for the given technology, so that most people were barely at the edge of subsistence. The Great Famine of 1315-25, which killed as many as one-tenth of the population, illustrates the precarious character of life for most of England's population in the early 1300s. The plague, however, killed greater numbers. After the first three major plague epidemics, England's population in the 1370s is estimated to have been 2.8 million and as late as the 1520s, it was still estimated to be only 2.3 million.

Although the initial impact of plague and its high death toll produced a depressed economy, after the shock of the terrible mortality diminished, the decrease in population ensured that fewer people lived at the very brink of subsistence. Gradually, as the remaining population began to take advantage of their improved economic status to produce goods, material wealth, and possessions increased. It was a slow process but ultimately it produced an increase in material wealth that fostered the development of a merchant class. Eventually, this increase in material wealth resulted in more people having the financial wherewithal to destroy potentially infective objects as recommended by the physician's advice published along with the plague orders.

After the plague entered England, most of the marginal lands were quickly returned to pasture and wasteland, and while some manors prospered and quickly returned to pre-plague tenancy levels, many did not. Colin Platt's research also demonstrates that at least in some regions the size of the average land transaction increased substantially in the 150 years after 1349. Despite laws drafted to control the movement of serfs and free peasants, it appears that tenants gravitated to manors where the combination of good soil, low rent, and low service demands offered the promise of the best returns.

Even during the initial outbreak of the plague, there is some evidence that plague morbidity and mortality rates were lower among the upper than the lowest classes. At the time of the initial plague outbreak, however, there were a great many more peasant workers than nobles and elite clergy, and further, many landholders lived only slightly better than their peasants. In 1348 there was a relatively small artisan class, compared to that which had developed by 1600, so the impact of wealth on mortality levels cannot be directly compared with data from the later epidemics. Records indicate that only a few of the upper-tier elite were infected or killed by the plague during the initial wave of the Second Pandemic. Princess Joan, one of the few royal victims to succumb to plague, died in August 1348, in Bordeaux, on her journey to Castille where she was to marry King Pedro.

The reason wealthy people appear to have been at less risk from plague is unclear, but possibly their large estates allowed them the opportunity for a secluded existence, despite their large household staff. Alternatively, their lifestyles may have reduced their environmental contact with rats and fleas. Pope Clement VI, the resident in Avignon, is reported to have spent much of the first outbreak more or less in isolation and surrounded by hot fires, an uninviting environment for rats. Additionally, and of greater statistical significance, evidence from British ecclesiastical clerical records indicates that during the first plague epidemic British bishops had a considerably lower mortality rate than did the regularly beneficed clergy.

Even so, evidence from plague epidemics before 1600, however, does little to suggest that the poor were substantially more susceptible to plague than those who were better off. Monks of Westminster, well off and well-fed compared to the population at large, seem to have experienced plague mortalities similar to those recorded among English parish priests, and thus, by extrapolation, similar to mortality rates of the population as a whole. In the 1600s, however, the data show that poor parishes were suffering higher plague mortalities than wealthier parishes, at least in the larger metropolitan areas. Paul Slack examined mortalities in London as well as Exeter, Bristol, and Norwich, and found that this differential applied in all these cities. In London, Slack found that the period between the epidemic of 1563 and 1603 marked a distinct change in the epidemic patterns. In the epidemics after 1603, it became evident that the outlying, poor parishes suffered the brunt of plague epidemics, while the central and more affluent parishes were less seriously affected. Although the parishes most heavily struck by plague had varied in each plague outbreak, after 1593 the central London parishes were never again the locus of the highest mortalities in London. Slack also noted that after 1563 the differentials between parishes most and least devastated by plague increased. During the last 100 years of plague in London, it became evident to contemporaneous observers that the poor were disproportionally affected. In the early sixteenth century, King Henry VIII showed what now seems an undue concern about the threat posed by the English sweating sickness, while he remained relatively unconcerned about the threat of plague. His concerns may have reflected the reality that although plague mortalities, in general, were much higher than those of the English Sweat, he was at greater risk of the sweat.

By the end of Queen Elizabeth's reign, Londoners believed that plague was uniquely terrible and universally dangerous, and yet they also knew that plague was a disease especially dangerous for the poor. Initially, this differential in morbidity and mortality between rich and poor was linked to bad humor, which collected in small stuffy rooms common with many inhabitants. Over time, several reasons for the differential mortality were considered. Defoe's primary explanation for this phenomenon was based on a theory of contagion. He says that during a plague epidemic, poor people were required by economic necessity to travel on city streets to continue performing their duties and thus were exposed to the infection because economic necessity forced them to engage in risky behavior. Defoe's basic advice for avoiding the plague was to stock up on material necessities and retreat from the world of daily activity in London. In light of our contemporary understanding of plague as a disease carried by the fleas on the backs of rats, the Elizabethan idea that plague lingered in the vapors of overcrowded stale rooms seems quaint, and neither of Defoe's explanations is entirely satisfying.

Nonetheless, there has been very little investigation into what might have produced the change in relative morbidity and mortality between London's rich and poor. First, it is important to note that while London was not by any means an egalitarian environment, neither was it as socially segregated as a contemporary urban environment. The rich and the poor lived in relatively close quarters. Servants and apprentices often lived with their masters, or in quarters provided by their masters, and people typically lived just behind or above the place where they worked. There is some evidence that during the early modern period servants increasingly came to live in separate apartments. Although the rich and poor lived in separate housing, individual city blocks included houses occupied by poor as well as rich; housing was often divided so that poor person lived on the less desirable side of the block.

Parishes were not all equally wealthy as is demonstrated by hearth tax records, but the population of individual parishes included both rich and poor. As the mortality and relative mortality levels of individual parishes shifted, population demographics were changing, but they did not change enough to explain the changes in mortality levels. Nonetheless, the increase in mortality among the poor in conjunction with a decrease in mortality among the rich suggests that changes in human living styles and behaviors could also have played a significant role in determining morbidity and that some combination of the many societal changes, which occurred during the early modern period, reduced the risk of the plague for the more affluent members of society.

Many of these changes are well known and often discussed, but not in the context of how they may have affected the distribution of plague, a topic I will now address.

Urban Construction

Construction techniques were constantly changing in London across the period of plague visitations; however, these changes were unlikely to have affected the access that rats had into houses and tenements. The urban setting was also dramatically altered by King Henry VIII's transformation of sacred landscape into secular and by the enormous increase in London's population. Between the fourteenth century and the seventeenth century, changes in housing construction were developed to reduce building costs. By the late Tudor period, building materials had become quite expensive within London because most had to be brought in from surrounding regions. The dimensions of some of the construction materials became standardized to facilitate prefabrication and some buildings were at least partially prefabricated outside London. This increase in uniformity may have had the result of making building components fit more tightly together, which could have hampered rat movement. However, as new construction alone would have been affected by these changes, they could have had only a minor effect citywide.

Also, changes that decreased the cost of construction by producing more uniform components were combined with construction changes that reduced the material requirements for construction. An early example is a use of arches in the undercrofts, or basements, of buildings. Also, as noted in chapter 4, shared foundation walls gradually came to be made two feet rather than three feet wide to reduce expensive stone building within London. With increasing costs, stone gradually came to be an insignificant component in the construction of houses. Solid stone walls were replaced by walls made either of brick or with stone exteriors filled with chalk. By 1610, when Ralph Treswell (c1540-1616) surveyed London houses, very few houses had walls of stone and none had four stone walls. Additionally, attempts to reduce construction costs were coupled with the addition of visual details. For example, in some late sixteenth-century buildings, closely placed studs were employed that were not joined to horizontal beams, and so were merely for visual effect. Such construction techniques may have left gaps for rat movements. Furthermore, interior walls were generally flimsy, "commonly lath and loam finished with a skim of plaster." They were permeable enough that in 1390 a burglary was committed by going through the wall of an inn. Such walls of porous lath, lacking a solid core and with irregularly placed or non-standard studs, would have moved by climbing rats simple and undetectable.

Throughout the Second Pandemic, one of the most important changes in London's built environment was that it came to house many more people. As described in chapter 4, London's population increased from about 50,000 in 1348 to about 141,000 in 1600 and then to as many as 459,000 people in 1665. With such an enormous and rapid population increase, housing the population posed a considerable problem. While much of this population increase occurred in the suburbs where house regulations were less stringent, available open space within the city was built upon and existing houses were subdivided into crowded tenements, some of which forced dozens of people to share one cesspit.

Many larger, wealthier households were

equipped with rudimentary plumbing, or piping, that connected necessary rooms, or latrines, to cesspits. These pipes or tubes were usually made of wood rather than stone, but in either case, their passage through house walls would have provided space for rats to move through and space into which dead rats could have disappeared. Changes in building construction in London could hardly have affected the number of rats living in human habitations and even twenty-first-century urban environments all over the world are rife with rats. In Egypt during an epidemic of the Modern Pandemic, bodies of many dead rats were found on the floor of a house of infected people, and even more, bodies were found within the structure of it when the house was partially demolished.

Twigg argues that evidence of this sort, as well as that from newly constructed Egyptian houses that became dangerously infected with rats and plague, indicates that changes in construction could have had no impact on the distribution or severity of the plague. The small plague outbreaks in early twentieth-century Britain, however, demonstrate that although plague-infected rats were distributed across a large region, only a few humans became infected. It is possible that while better construction cannot control rats, better construction, especially solid walls, may separate rats from humans just enough to make the transmission of plague from rats to humans much less likely. Even though contemporary cities are rife with rats, humans are rarely exposed to plague even where plague is present because rat control and current construction methods, tend to keep humans and rats separated.

As London grew, its emphasis on markets increased, and its main streets were widened. Also, to facilitate the movement of wagons and riders, regulations that controlled the height and extent of jetty overhangs were strengthened and enforced.

Suburban areas, aside from the main roads into town, did not share London's emphasis on providing open space for traffic and markets, although merchants used the space just outside the gates as informal markets to avoid paying London's entrance fees on merchandise. Contemporary experience suggests that rats are likely to congregate around granaries and to travel with food shipments, so these markets outside the gates could have played a role in the spread of the plague. However, within the walls, London epidemics do not seem to have centered around ports where foodstuffs arrived, or around markets. Cornhill (74, 90) parishes suffered extensively in 1563 and 1593, and again in 1625; however, by 1625 other parishes were hit substantially harder than these in the heart of central London. In 1665 Samuel Pepys noted that St Michael Cornhill was one of the least devastated parishes. It is possible that widening streets to accommodate market traffic and to allow for market development made market spaces less appealing to rats and kept the rats of individual housing blocks more isolated one from another.

Religion

In the initial wave of the Second Pandemic, pestilence was seen almost universally as a sign of God's displeasure. English chroniclers noted that plague seemed to bring out the worst in people, and the *Historia Roffensis* notes that "those who once had to work now have time for idleness, thieving and other outrages." John of Reading notes that although many people died and left all of their worldly goods behind, "all these worldly goods were not enough for those few who remained alive." He also wrote that this new-found wealth "wounded the regular clergy very much, but wounded the mendicants fatally." So even the redistribution and temporary affluence produced by the high death tolls were seen by some as a misfortune.

The plague was described as a punishment meted out on society as a whole, not against individuals, however, and so people publicly and communally attempted to appease God and thus to stop the disease. In the early years of the pandemic, religion had played a very active role in the battle to combat the plague. On the continent, there were many public religious events, both sanctioned and unsanctioned, such as wild gatherings of flagellants. When the plague first arrived in England, the response seems to have been less disorderly, but public religious observance was seen as a primary defense against the disease. In the first years of the pandemic, bishops organized processions and large-scale masses to be said in an attempt to defend the country against the onslaught of plague. By the seventeenth century, this mindset had changed. Although God's favor continued to be seen as necessary for health, by the seventeenth century public religious activity was a less significant aspect of society's battle against the plague than it had been in the early years of the pandemic. During the last century of plague in London, efforts at improving urban sanitation, involving individuals and government working together, had taken on a significantly larger role in combating plague, and public religious expression had become less important.

During the 300 years of plague epidemics in England and London, it is clear that the role of religion and religious activity changed with respect to plague, but this does not mean that plague ceased to be seen as a mark of God's disfavor. For example, in 1625 Thomas Dekker (c.1570-1652) wrote that "God will not haue his Strokes hidden: his markes must be seen." In other words, it was useless to attempt to dissemble by hiding the true nature of deaths caused by plague, because God's message could not be hidden. A generation earlier, Kellwaye included the following statement in the dedication to his patron:

> *Wherein God hath alreadie drawne his sword against us, and striken some fewe, and except we cause him by our speedie repentance to sheath it, he (no doubt) hath determined to strike us at the quicke, how fearfully the wrath of God consumes, if his indignation be once kindled.*

Further, in the introduction to his work, Kellwaye emphatically noted that pestilence "is a just punishment of God layde upon us, for our manyfold sinnes and transgressions against his divine Majestie." In Chapter 4, "Sheweth how to prevent the plague,"

Kellwaye's first suggestion is to acknowledge one's sins and wickedness to God.

Despite Dekker's fervor, the reliance on public worship as a means of combating plague had dwindled. In 1625 the House of Commons Journal records a substantial battle among its members concerning establishing a day of fasting as a public religious observance. On June 21, 1625, the House ordered a day of fasting for House members before they took communion as a religious act in response to the plague epidemic and three other issues of public concern. One of the arguments posed by opponents of the decree for a public fast was that it should not be broached until the House members' fast had been accomplished and this point of view won out. The House of Commons ordered a day of fasting only after they had performed their own. A majority of the members of the House of Commons felt that they should put their own house in order before urging the country to prayer. Also, this reticence suggests that they did not think public fasting was likely to produce a direct response from God. Thus, in July, churchwardens were finally told to "exhort their parishioners to observe the public fast ... privately in their own homes."

The request that the fast be observed privately at home, rather than publicly in Church, avoids the issue of what churches are appropriate for religious service performed as a civic duty, and it also alleviates concerns any of the house members may have had about contagion during services. After July 20, these fasts were held on successive Wednesdays. The first of these weekly fasts was celebrated with a public service at Westminster for the King and Lords, which served as a collection point for mandatory donations for the assistance of people devastated by the plague. Thus, the fasts took on the characteristics of a social effort to help people, rather than a religious observance to exhort help from God. In 1665, special religious observances were again decreed but the public observation was relegated to a monthly fast with prayers.

Although special services to appease God and eliminate the plague became less significant, church worship and church services were central to both public and private life in early modern London. Even though church services were not restricted during plague outbreaks, attending worship was not seen as an entirely risk-free endeavor. In 1592, rules stated that nobody who had died of plague was allowed to remain in "any church in the tyme of Comon Prayor, Sermon or Lecture." Nonetheless, because church services, held in parish churches, were regarded as local experiences rather than public gatherings, they perhaps were not seen as public and dangerous in the same way that entertainment events were. London churches were primarily local institutions as exemplified by the fact that the parish served as the smallest political division and the distributor of public welfare assistance.

English regulations not only restricted the role of public religious observance, but they also limited the possibility of individual charitable, or Christian, visits to the sick. After 1578, the English government began legislating the separation of people with plague or those who had been exposed to plague from those who were free of the disease. Only authorized people, who tended to be impoverished members of the working class, were allowed into the houses of the sick; thus, the care of the sick became a public rather than a personal religious duty. The Plague Orders of 1578, which were repeatedly reissued, make the shift from private to public care explicit. The regulations spell out the taxes that are to be collected and what sort of people are to be hired to guard the sick, to help the poor and to investigate reports of sick people:

> *If there be any person, Ecclesiastical or laye, that shall holde and publishe any opinions (as in some places report is made) that it is a vayne thing to forbeare to resort to the infected, or that it is not charitable to forbid the same pretending that no person shall dye but at their tyme prefixed, such persons shalbe not onely reprehended, but by order of the Bishop, if they be ecclesiastical, shalbe forbidden to preache, and being laye, shalbe also enjoyned to forbeare to utter such dangerous*

opinions upon payne of imprisonment, which shall be executed if they shall persever in that error. And yet it shall appeare manifestly by these orders, that according to Christian charitie no persons of the meanest Degree shalbe left without succor and reliefe.

That the concept that Christian charity could be best achieved by limiting the freedoms of plague victims remained an integral aspect of the plague orders, although as described in Simon Kellwaye's 1593 treatise *A Defensative against the Plague*, this concept does not seem to meet with his approval. This is one point on which Kellwaye's suggestions vary slightly from officially sanctioned behavior. Chapter 12 of Kellwaye's first treatise addresses what people should do to protect themselves when they visit the ill, which indicates that he does not believe that plague victims should be cut off from all visitors. Although the visits that Kellwaye describes are medical, they do not appear to fit within the limitations of a visit by searchers or a public surgeon, which the plague regulations prescribe. The care-giving Kellwaye describes requires a virtually 24 hour-a-day commitment. Although some of Kellwaye's medical advice seems useless, and some are potentially dangerous, for example, his insistence on emetics, the continual care that he recommends might have increased the likelihood that patients would survive the plague. The relatively good palliative care given to Europeans in outbreaks of plague in India during the Modern Pandemic explains their relatively lower mortality levels. Kellwaye also suggests, as do the plague orders, that people who have been exposed to the plague by assisting victims should carry some marker when they walk in the streets so that they can be avoided.

Thus, while religious observances remained central to society in the seventeenth century, public religious observance was no longer seen as an effective means of combating the public menace of plague, even though the disease was still seen as a manifestation of God's displeasure. The locus of plague response passed from the realm of religion to the realm of civil authorities, although the ultimate cause of the scourge was still seen to be in God's control.

Public Gatherings

Church services, unlike most other gatherings, were not sharply curtailed during plague outbreaks. Plague orders placed a priority on controlling rowdy gatherings, especially events targeted to the lower classes and activities where people from diverse social milieus might mingle. Pepys mentions being surprised at the social diversity he saw at these public events. By the late sixteenth century, most normal public gatherings including guild meetings, theater events, and entertainments, as well as Parliament meetings, were canceled or forced to move outside city limits during plague epidemics. The plague orders of 1636 forbid all public feasting "particularly by the Companies of this city" as well as any feasting in any other establishments for public dining. The orders even go so far as to suggest that money saved by this restraint could be "imployed for the benefit and relief of the poor visited with infection." Orders specifically prohibited all "Playes, Bearbaitings, Games, Singing of Ballads, Buckler- play or such like" that attract crowds of people. Pepys' characterized the audience of cockfights at Shoe Lane, as a "strange variety of people from Parliament man ... to the poorest prentices, bakers, brewers, butchers, draymen and what not." Thus, his description of these audiences as composed of a broad cross-section of society may suggest why these events would have been seen as more public and therefore more dangerous than attendance at a local parish church.

Funeral processions, as well as any gathering at the houses of the deceased, were forbidden. For example,

> *that the buriall of the dead by this visitation bee at most convenient houres, allwayes eyther before Sun-rising, or Sunne-setting, with the privity of the Churchwardens or Constables, and not otherwise and that no neighbours nor Friends be suffered to accompany the Coarse to Church ore enter the house visited.*

The orders additionally prohibited a corpse of a person who died of plague to be in a church during "Common-Prayer, Sermon or Lecture." So, although the orders do not restrict religious gatherings, they did attempt to reduce the risk of church services by eliminating the possibility that a plague-infected body would be present during the services. While many of the orders require that the authorities, aldermen, and other officials within London meet to arrange for enforcement of the orders, most of them stipulated that these meetings should be convened in areas free of plague. Moreover, the orders also stipulated that only officials living in areas free of plague should attend these meetings. The regulations controlling public gatherings suggest that plague was seen to spread in some manner from person to person, but they also suggest that disorderly gatherings where strangers met were seen as more dangerous than local gatherings.

Sanitation, Garbage, and Cleanliness

Virtually from the first plague epidemics, one of the consistent responses was to order a clean-up of the urban environment. Edicts and orders to clean up the City, to remove filth, human waste, and garbage, were issued to combat each of the plague epidemics. These efforts to clean up filth, especially urban filth, seem to have been driven by the general understanding that odors were integrally related to the spread of disease. Thus, anything that reduced odors could be defined as cleaning. Further, people used sweet-smelling herbs not merely to avoid experiencing bad smells, but also to block the absorption of foul odors. However, it was not only bad smells but also rot and putrification that were linked with disease. Nevertheless, the orders had to be reissued with each plague outbreak, which suggests that in the interim between plague outbreaks, cleanliness regulations were not uniformly followed or enforced.

Nonetheless, over the centuries, demand for improved urban sanitation and the removal of foul-smelling substances remained a consistent response to plague epidemics.

In 1349, King Edward III complained to London authorities about the "human feces and other obnoxious filth lying about in the streets and lanes." While it is possible that the King's complaint was part of an ongoing struggle between the crown and city officials about the control of London, or that it provides evidence that basic city services had been disrupted during the chaos of the plague epidemic, this complaint likely indicates that during pestilential times noxious odors were perceived as more dangerous and threatening than in ordinary times. In 1361, during the second plague epidemic, Edward III sent out a writ that demonstrates that by the second major plague outbreak, the King's concern with the sanitation of London had expanded beyond a concern for human wastes:

> *Because by the killing of great beasts, from whose putrid blood running down the streets and the bowels cast in the Thames, the air in the city is very much corrupted and infected, hence abominable and most filthy stench proceeds,*

sickness and many other evils have happened to such as have abode in the said city, or have resorted to it: great dangers are feared to fall out for the time to come, unless remedy be presently made against it. We, willing to prevent such dangers, ordain by consent of the present Parliament that all bulls, oxen, hogs and other gross creatures be killed at either Stratford or Knightsbridge.

Despite the King's pronouncement, the dumping of both household and commercial wastes, such as those from slaughtering animals and tanning hides, continued to be a problem within London for centuries. Kellwaye also mentions another dangerous, but usually overlooked waste product: blood. He states "that no Chirurgions, or barbers, which use to let blood, doe cast the same into stréetes or ryvers."

Before 1383 it was not legal to place a latrine so that it dumped into a moat or stream flowing through London, but complaints before this date suggest that it had been a common practice. The 1383 legalization of latrines over waterways was accompanied by an annual fee assessed on all who maintained such privies, a fee which was to be used to keep the stream free of debris and blockages, to ensure that the human wastes were carried down to the Thames. Also, the regulations forbade householders to dump other household rubbish into the streams. Despite these regulations, there were sporadic complaints that debris and human wastes produced blockages so much that the waterways began to stink. These problems occurred not only in the Fleet but also in the Walbrook and the moat or ditch abutting the exterior of the London wall. Complaints filed in Assize de Nuissaince and through other venues suggest that fighting filth in London was an ongoing battle.

In 1462, London prohibited latrines over the Walbrook and the Fleet and ordered that landholder along with the Walbrook cover it over to ensure that it could no longer be used for sewage. In 1477, the city extended rules against latrines emptying into the Walbrook to include other urban ditches and moats. Nonetheless, the practice of emptying latrines into riverways appears to have continued.

The 1578 plague orders and most subsequent orders issued in response to plague outbreaks placed considerable emphasis on keeping streets clean. The London orders enumerate several cleaning requirements for London citizens. Every householder with a pump or a well was required to pour ten buckets of water down the gutters in the street before six in the morning, and another ten buckets of water down the gutter after eight in the evening. These Orders further enjoin people from pouring water into the gutter in a way that washed material from the gutter into the street. Also, householders were required to sweep the mud and filth of the street out of the gutters when they were rinsed out, so that debris did not pile up and block the canals.

In the final step of the street cleaning process, scavengers and rakers were required to clean the streets of large debris and piles of garbage every day except Sunday. Regulations, issued by the Court of Common Council, also required that street pavements should be kept free of "hooles ... wherein any water or filth may stand to encrease corruption or Infeccon." While the London orders were very similar to the

Royal orders, the Royal orders did not address urban issues such as street paving. The proceedings also forbid the accumulation of dunghills outside "of stables, brewhouses, or other places." Dung piles were forbidden both in streets and in other open spaces, not only in London but in the surrounding suburbs as well, upon pain of imprisonment.

Plague orders issued by London in 1608 emphasized the need for cleanliness during times of plague. They required householders to sweep the street in front of their doors daily, and rakers to remove the resulting debris daily. The orders further stipulated that "the Raker shall give notice of his coming by the blowing of a Horn as heretofore hath been done." An additional measure of cleanliness that this order required was that laystalls be removed as far as possible from the city and away from common passages. It also specified that "Nightmen" should not empty a "vault into any Garden near about the City." Presumably, although only night workers were singled out under this provision, dumping human waste at any time was not acceptable.

Plague orders also proscribe the selling of

unwholesome fish, flesh, musty grain, and rotten fruits of any sort. Brewers and tippling-houses were singled out for special attention as the orders specified that their casks should be checked to assure that they are not musty or unwholesome. Additionally, the rules stipulated that tippling houses that allowed disorderly drinking would be "severely looked" into. The regulations also forbid any "Company or person to be suffered to remain or come into any Tavern, Alehouse or Coffee-house to drink after nine of the clocks in the Evening." Presumably, selling rotten or unwholesome foods and drinks was generally frowned upon, but rules concerning this sort of abuse appear to have been more stringently enforced during plague epidemics.

It was not only the political entities issuing plague orders that indicated that cleanliness was considered an important aspect of combating the plague. *Present Remedies against the plague,* published in 1594 and "written by a learned Physition for the health of his country" expresses a similar concern. The author begins with a brief message "To the Reader" that notes that the "infection of the ordinary disease called the Plague or Pestilence" is again on the rise so he has decided to publish advice for dealing with the problem. His recommendations begin with his advice on cleanliness.

> *Right necessary and convenient it was, that you keep your houses, streets, yardes, backesides, sinks, and kennels sweet and clean from all dangling puddles, dunghills, and corrupt maynures which ingender stinking sandurs that may be noysome or breed infection.*

Kellwaye also includes cleanliness as an essential aspect of 'order,' which he sees as an important aspect of limiting plague outbreaks. The first two of Kellwaye's "three principall meanes" of preventing plague are first to pray and admit one's sins, and second to flee plague infected areas. His third means for avoiding plague consists of three points: "Order, Diet and Physicall helpes." Under the rubric of 'Order,' Kellwaye addresses the importance of cleanliness; however, for Kellwaye, clean seems to have been virtually synonymous with smelling sweet as he refers to keeping the house "cleane and sweete." In this short section of his treatise, Kellwaye recommends "not suffering any foule t [and] filthy clothes or stincking thinges to remaine in, nor about" your room, and opening the north-facing and east-facing windows in the daytime "if the aire be cleare and that no infected and unsavory" smells are near the window.

London was not the only town concerned with urban cleaning. In Haverfordwest during the epidemic of 1652, public buildings were lime washed in an attempt to disinfect them. This cleaning occurred despite difficulties the town administration was facing to ensure that all the impoverished sick inhabitants were fed. Haverfordwest was experiencing economic hardship because of previous over taxation and because many of the wealthier tax payers had left town due to plague.

Because the idea of cleanliness was inseparable from that of smelling sweet, the process of cleaning a house, especially in the case of cleaning to eliminate plague, frequently included making a fire and burning sweet smelling herbs. Indeed, it is often difficult to tell the extent to which burning sweet smelling herbs was not the primary aspect to cleaning. Since disease was associated with vapors, smog, damp, and even sweat, it may have been that burning a fire, drying the rooms out, and fumigating them with smoke from sweet smelling herbs was considered the most important aspect of cleaning to prevent plague. Nonetheless, the many regulations and complaints about abuses suggests that at least during plague epidemics people made an effort to improve sanitation to combat plague.

Cloth, Clothing, Fibers

From the earliest plague outbreaks, the re-use and distribution of clothing and fibers that had been used by people infected with plague were understood to have been inextricably connected to the spread of plague. Boccaccio noted that people were infected with the disease by many means, including airing clothes for the ill or simply touching the clothes of an ill person. He noted that many people said that the clothes and other possessions of those who died of the contagious pestilence could bring death not only to other people but to animals, including dogs and cats. He also wrote of seeing two pigs die after rooting around the ragged linen and woollen clothes of a plague victim. Nevertheless, despite the fear of contagion associated with used clothing and bedding, because fabric was relatively expensive, it was extensively reused. Plague orders stipulated that items of little value should be burned and that bedding and clothes from plague victims had to be cleaned and well aired before it could be sold. Further, fabrics, bedding, clothing and cushions often formed an important segment of property bequeathed in wills.

London plague orders of 1625 begin by specifying that while a house is under quarantine "no Clothes, linnen or other likething be hanged out or ouer into the streete," and conclude by describing how houses should be aired after the quarantine is over. "No clothes or other things about the infected be given or sold, but either destroyed or well and sufficiently purified. On paine of punishment." Although Kellwaye's treatise devotes only a few words to material items that are not medicinal, he mentions that coming into contact with a person who has recently had plague is a common way of contracting the disease but that "for the most part it doth come by receaving into our custody some clothes or such like things that have been used about some infected body." As previously mentioned, Kellwaye considered removing stinking clothes as a preventative measure against plague.

The London plague orders issued in 1665, although substantially reworked, begin by referring back to the orders issued "in the first year of the Reign of our late Sovereign King James of happy memory." They include several regulations that demonstrate concern about the dangers posed by fabrics. Under the heading "Airing the Stuff," these orders require the

> *sequestration of the goods and stuff of the infected, their Bedding and Apparel, and Hangings of Chambers must be well aired with fire, and such perfumes as are requisite within the infected House, before they are taken again to use: this to be done by the appointment of the Examiner.*

These Orders further placed restrictions on vendors of used materials. Under the heading "No infected Stuff to be uttered," the orders required that "no Brokers of Bedding or old Apparel be permitted to make any outward Shew, or hang forth on their Stalls Shopboards or Windows toward any Street, Lane, Common-way or Passage any old Bedding or Apparel to be sold." Further, the regulations stipulated that vendors who removed "any Bedding, Apparel, or other Stuff out of any Infected house, within two months after the infection hath been there shall have their own houses shut up for 20 days." These requirements, coupled with the stories recounted by various observers, strongly suggest that fabrics were considered of special concern.

Royal plague orders issued in 1578, 1592, 1593,1603, 1608, 1625 and 1629, are virtually identical aside from the introductory materials, which include references to current events. These orders were printed in conjunction with medical advice from "the best-learned physicke in the realm." Printing the laws about behavior during plague epidemics along with medical advice suggests that these pamphlets, including the laws, were intended to be of actual benefit to individuals as well as to the overall society. Although the advice from the physicians does focus on medical preventatives and treatments, it also includes brief warnings about issues of sanitation. It opens with information on perfuming houses, which is followed by instructions, under the header "Perfuming of Apparel," for sanitizing clothes. The physician recommends that frequently worn clothes should be kept clean and scented and that as soon as possible after exposure to plague, clothing should be aired. In the final section of the medical advice, under the heading "Infected Clothes," the learned physician states that contagion is suspected to remain in both woolen and linen, and that the best means of disinfecting clothing is "fire and water." The author further recommends that clothing be washed often and aired in the sunshine, whether under cold or warm conditions, and that clothing of little value should be burnt to control the spread of the plague. This advice or its equivalent was repeatedly reiterated throughout the pandemic. Over time, more people must have read or heard this advice, and as material wealth increased more people could afford to heed it.

Pepys recounts a story about a man in a house under quarantine who attempted to save the life of his last little child from certain death that he felt would have been its end had it remained locked in their home. In this story, the man passes his naked child through a window into the arms of a waiting friend. The story not only illustrates people's concerns about the dangers of closing people up in infected houses, but it also illustrates the special concerns that were attached to cloth and fiber. It indicates that the baby's clothing was considered to present a greater danger of transmitting the plague than did the baby.

During outbreaks of the Modern Pandemic, inspections have been conducted on people and their clothing to see whether fleas were carried along with people or their clothing. The results of these inspections suggest that fleas are not commonly transported in clothing. However, before concluding that it was unlikely that clothing and fabrics transported fleas, and thus plague infections in the past, it is important to consider the condition of fabrics of the past. The cloth was an expensive commodity and was extensively reused. Clothes were not frequently washed, despite the recommendations of medical authorities, and were worn hard for many years. As fabric, clothing, and household items deteriorated with use, rather than being thrown away, they were simply relegated into the service of a person of lower status. Under these circumstances, clothing could reasonably be considered a reservoir of an unknown contagious substance. It is also interesting that during a small outbreak in Glasgow between 1900 and 1910, rag collectors were among the few victims.

Dog Catching

Amongst the behaviors and activities adopted throughout plague epidemics in early modern England that may have influenced the outcomes of epidemics is the wide-scale destruction of numerous vermin within various cities. Although plague orders recommend the destruction of many potentially dangerous animals, the primary animals singled out for destruction were loose and annoying dogs.

In 1563, the orders from the Lord Mayor of London "commanded that no dog be allowed out of any house without a lead on pain of a 3*s* 4*d* fine for the owner and death for the dog." The orders in 1592 were even more stringent; the mayor gave the "common huntsman ... special charge to kill every dog or Bitch, as shall be found loose in any street or lane" or howling or otherwise annoying the neighbors, and these orders include the penalty that any huntsman who is negligent or shows favors and spares dogs will lose "his place and service, and suffer Imprisonmente." The1603 plague outbreak similarly produced the slaughtering of many dogs, and in 1636 "Orders for Health" said:

no Hogs, Dogs, or cats or tame pigeons or Conies be suffered to be kept within any part of the City, or any Swine to be, or stray within the Streets or Lanes, but that such Swine bee Impounded by the Beadle or any other Officer of the Common Counncell, and that the Dogs be killed by the Dog-killers appointed for that purpose.

Orders demanding the killing of dogs were not limited to the city of London. Similar regulations were also occasionally ordered in cities on the continent as well as in parishes surrounding London, including St. Margaret's in Westminster and St. Martin-in-the fields. During the epidemic of 1592-1593, St. Martin-in the-Fields hired a dog catcher to kill the dogs of their parish. During the 1603 plague, the killing of 502 dogs was paid for in St. Margaret Westminster, and in 1625 this parish paid £2 17*s* 8*d* to kill 466 dogs. Laws that required the death of loose dogs were common in many cities in England and were also issued in Scotland.

Dogs were slaughtered in years with only minimal plague outbreaks, but the numbers killed were much higher in years of heavy plague mortalities. According to London Chamber accounts for the years from 1584 to 1586, payment was made for the killing of 1,882 dogs, which contrasts dramatically with the number of dogs killed during the summer plague outbreak in 1636 when payment was made for the death of 3,720 dogs. Throughout the final Great Plague of London, at least 4,380 animals were killed. Dogs were seen as dangerous, at least in part because they were disruptive. Regulations specify that both loose dogs and those "within their doores, making howling or other annoyaunce to their neyghbours" were subject to elimination. The author of *Present Remedies* also warns that because dogs "runne from place to place, and from one house to another," and get into unclean things in the street, they can bring infection into the house. This would seem to indicate both a fear of contagion as well as a fear of dogs' disruptive behavior. The orders also require the elimination of animals other than dogs within city limits; however, the payment records do not always specify for what animal deaths a dog catcher was being paid. Kellwaye warns that plague can be brought by "dogs, cats, pigs, and weasells which are prone and apt to receive and carrie the infection from place to place."

Pepys' Diary entry for August 21, 1665, expresses concern about walking late at night past the isolated Coome farm because he is afraid of dogs, rogues, and plague which have infested the farm. He then notes that it is odd that such an isolated farm was infested with plague, but attributes the contagion to the fact the landholders allowed beggars to sleep in the barn. Although he does not explicitly connect those fears, mentioning them all in one sentence implies a connection. The entry for August 12, which makes the baseless claim that the Lord Mayor has ordered people inside by nine at night so that "the sick may have the liberty to go abroad for ayre," suggests that Pepys saw a connection between night, disorder and disease. Indeed, traditionally, some noisome and onerous tasks had been relegated to the night hours.

It is difficult to determine exactly what the effect that killing urban vermin might have had on the spread of plague and the distribution of epidemics because the effect that might be expected depends on many factors. To predict the effect of these attempts to rid London of vermin, it is important to know not only the number of animals killed but also what kinds of animals were being killed, and when. If only dogs were killed, it seems likely that the rat population would have exploded, especially if they were being killed between periods of epidemic plague. Dogs not only hunted rats but also competed with them for urban garbage. Mark Jenner provides evidence that dogs were being killed even during periods of relatively low plague mortality. If the rat population was kept relatively low, killing vermin probably reduced the frequency of epidemics. On the other hand, killing rats primarily during an epidemic might have intensified the human plague experience by driving *Y. pestis* infected fleas to people. Additionally, occasionally individual rat catchers probably would have been infected by coming into direct contact with infected rats and their fleas.

The killing of even very large numbers of rats, regularly, would probably have only a limited effect on the severity of plague outbreaks; however, it might have affected the locus of plague outbreaks. For example, if wealthier house holders-controlled rats more effectively than those living in tenements, it is possible that rat control was a factor in the shift of plague loci from the center of town to the suburbs. However, without extensive and continual use of poison, it is unlikely that enough rats could be killed to significantly reduce the risk of plague epidemics.

During the final century of English plague epidemics, the use of poisons did become more common and it is possible that mortality differentials between various areas in London increased in the seventeenth century because those who could afford to poison rats did. If, on the other hand, the killing of large numbers of rats was triggered by the onset of a plague epidemic, the result could be that fleas potentially already infested by *Y. pestis* was deprived of rat hosts and thus driven to human hosts. In this scenario, it is possible that the killing of large numbers of rats could intensify human mortality levels during an individual plague epidemic. The issue of vermin control is complicated because in early modern

England, the exterminators' primary focus was on killing dogs, not rats, and although cats and other vermin also were official targets, it is not possible to tell how many of each animal was killed.

In the 1665 epidemic, plague orders not only mandated that domesticated animals be controlled, they limited Londoners ability to keep domestic animals, such as pigs, pigeons, and rabbits along with cats and dogs. In a letter to his clerk, dated July 5, 1665, Sir Robert Long (*d.*1673) expressed concern about the plague's effect on both his clerk and his household. Sir Robert instructed the clerk to "take all course you can agaynst the ratts, and take care of the cats: the little ones that will not tirre out may be kept, the great ones must be kild or sent away." Evidence suggests that dog slaughters were instigated at the beginning of plague epidemics. Although issuing orders for the control of loose dogs may have been a normal aspect of urban management, the fact that plague orders repeatedly reissue the mandate to deal with loose dogs suggests that officials associated dogs with plague. However, as Mark S. R. Jenner points out, wandering dogs, like wandering people, were primarily seen as a threat to the proper social order and only secondarily seen as dangerous because of their potential for spreading plague. It is also worth noting that as people fled London and died of the plague, there were likely more loose dogs, who thus would pose a greater nuisance than they did during normal periods. In 1603, the Venetian Secretary in England, Giovanni Carlo Scaramelli, wrote in a report home that "no steps have been taken as yet, except to kill the dogs and mark the houses." This statement suggests that killing dogs during the plague epidemic was not simply a response to an increase in wandering dogs.

Flight

From the time when the plague first arrived in Europe, flight from the area of plague was a common method of attempting to avoid the pestilence. An aphorism from an early plague tract that had been borrowed from even earlier sources suggested "flee quickly, go far, and stay away a long time," as the best method for avoiding the plague. The incredibly rapid diffusion of plague across all of Europe can only be explained by the flight of large numbers of people accompanied by rats and/or fleas. Throughout the plague's residence in England, flight remained a preferred method of evading plague.

Kellwaye's second means for avoiding contagion by pestilence, after prayer, his first suggestion, was to flee far, the farther the better; he further recommends not returning quickly for "feare of an afterclap." One of the explanations for an upswing in plague mortality toward the conclusion of the 1665 epidemic was the return of people who had fled during the height of the epidemic. Kellwaye reinforced the validity of the advice to flee by ascribing the advice to both Rondoletius and Valetius. He also points out, however, "yet were it a very uncharitable course that all which are of abillyte should so doe: for then how should the poore be relieved, and good orders observed." In 1625, Thomas Dekker made this argument even more emphatically. "*To the* Run-awaies *from* London you flye to save your selves, and in-flight undo others."

The evidence of widespread mortalities during the first epidemic wave suggests that flight was of little use at that time, and yet by the time of the Great Plague of London, the flight seems to have been a very successful means of avoiding the plague.

Throughout Pepys' descriptions of the rising plague mortalities of July and August are remarks concerning London's empty streets, and people who are leaving or who have left London for the safety of the countryside. He also mentions lines of traffic leaving London by Cripplegate, which suggests that it was the wealthy who were leaving and that they were leaving in carriages or carts, not on foot. This is important because evidence from the Modern Pandemic suggests that exertion after the infection is a contributing factor in producing and spreading pneumonic plague. Additionally, it is clear from Pepys'

Diary that those who could, began to leave London even while the epidemic remained primarily focused on a relatively few of the poorer parishes. Pepys' Diary, however, also makes it clear that people continued to leave London throughout the epidemic as the intensity of plague increased, and yet the disease does not seem to have followed those who fled into the countryside surrounding London. Eventually, the plague did become dispersed in the surrounding regions, which experienced sporadic epidemics through 1666, although they were not as intense as those in London.

Pepys' behavior during the plague epidemic provides yet more evidence that the plague was seen both as a disease of place and a contagious disease. The evidence he provides also suggests that the plague was not very contagious. In June, Pepys sent first his mother and then his wife away from London for their safety. Despite the danger London presented, Pepys continued to live an active life in London; furthermore, he made regular trips down the river to visit his wife in Woolwich. Still, Pepys occasionally mentions that he had come into contact with people who subsequently became ill or died, for example, cabmen and watermen. Nonetheless, Pepys seems to have had little concern that he would expose his wife to the plague. He did know, however, that other people perceived his journeys to London as a threat; when Pepys visited friends he was "forced to say to say that I lived wholly at Woolwich." Like Pepys, John Evelyn sent his wife out of London to escape the threat of plague, and also like Pepys, Evelyn visited his wife in her refuge away from the dangers of London. Neither of these men seem to have been especially careful or concerned about carrying plague to their wives. Admittedly the evidence in Pepys' Diary suggests that well off people had a lower risk of contracting plague than did laborers but it does not seem to be exclusively the result of their flight from London.

The various attempts to control the spread of plague suggest seemingly conflicting views of how the plague was spread, and thus these apparent conflicts might provide information on the early modern understanding of pestilence. A major issue is whether the plague spreads person to person or is associated with particular places. Cohn has used the speed with which the Black Death spread across Europe and much of Asia to argue that it, unlike modern bubonic plague which is caused by *Y. pestis*, was not a disease of specific locations. In the investigation of the Modern Pandemic, researchers began with the idea of plague as a contagious disease, whereas people of the afflicted regions interpreted it as a disease of location; eventually, researchers came to accept local perceptions that plague is a localized phenomenon.

During the last 100 years of the Second Pandemic, London was viewed as a central locus of plague so that flight, even flight of only a short distance beyond the city, was seen to be, and was an excellent precaution.

Public Health: Quarantines and Restrictions on Vagabonds

The policy of attempting to separate the well from the sick was an element of the public response almost from the plague's first appearance in Europe. The use of various forms of quarantine, or methods of attempting to segregate plague-infected people from the uninfected, was the one at least quasi-medical action that often had a direct influence on the morbidity and mortality rates of a plague epidemic. The effect produced, however, was not necessarily the desired one of limiting plague induced mortalities. Quarantining people within infected houses may have increased mortality and morbidity rates because sequestering people in infected houses assured their exposure to infected rats and fleas, and also reduced plague victims' access to palliative attention.

Efforts to segregate the population were practiced on what can be described as micro and macro levels; on the micro-level, people from houses known to be infected by the plague were restricted to their homes, often forcibly. On the macro level, travel was curtailed, either into or out of an area such as a parish, city, or country. Incoming ships often were isolated for a period of up to 40 days to demonstrate that the contents, both people and goods, were free of disease. At the macro level, it is possible that quarantines had some positive effect, although it is very difficult to control the movement of rats.

In Britain, quarantines were not employed with consistency until relatively late into the period of the Second Pandemic. Because the city-states of the Italian peninsula were densely populated, physically small, relatively well organized, and wealthy, they instituted various forms of quarantines much earlier in the cycle. By 1348, both Milan and Venice were employing quarantine measures with apparently very different degrees of success. Milan employed a harsh form of segregation that entailed simply shutting up houses where people were known to be sick and strictly controlling access into the city. Venice employed a more selective set of protective measures, including sending infected people to isolated islands and imposing "a general quarantine of up to 40 days... on all incoming ships." Despite Venice's theoretically effective quarantine measures, it suffered very heavy mortalities. During the first year of the pandemic, the mortality rate in Venice is estimated to have been approximately 60 percent, while the mortality rate in Milan, which employed draconian isolation techniques, had a mortality rate of only about 15 percent, which must have been one of the lowest in Europe. These examples point out the difficulty in analyzing the effects of the civic intervention on individual epidemic outbreaks. Although the records provide us with some information about control, they provide much less information about how successfully they were enforced, and they provide no information about the rat and flea populations or other ecological or social conditions that could affect the transmission of plague.

It was not until 1518 that England began to institute plague regulations, and it was not until 1578 that segregation of people in households with sick people, from healthy people, became the country's official policy. The earliest attempts at controlling the spread of plague were local and their focus was primarily to deny ill people, often defined as strangers, from entering a town or locality. Although Britain was relatively late to employ coordinated quarantine measures, individual cities enacted piecemeal regulations much earlier. For example, in 1349, Gloucester refused entry to travelers from Bristol, which was known to be infested with plague." Flight away from plague was always a recommended means of avoiding plague and people who could afford to isolate themselves did. Pope Clement VI is reported to have stayed in isolation and later, in England, King Henry VIII moved his household multiple times in an attempt to isolate his entire household from the infection.

Elizabeth I left London and instituted draconian measures to assure that no one in her entourage was exposed to anyone from London and thus who might have been exposed to plague. Those who could isolate themselves from the threat of contact with the sick did. People such as searchers, plague surgeons, and gravediggers who, due to their work, were known to have been exposed to the plague, were separated from the general population by being required to carry red or white staves to mark themselves as having been exposed to plague. Additionally, these workers who were regularly exposed to plague were expected to walk along the dirty channel side of the street.

In London, the closing and locking up of houses was spelled out in plague orders periodically reissued by the Crown and the London corporation, and reprinted with slight variations at times when the threat of plague resurfaced in London. The plague orders required that if one person in a house was found to have the plague, the house was to be closed with all of its occupants inside, guarded by a day and a night watchman, marked with a cross or an x, and have in large lettering the inscription "Lord have mercy" on the door. The orders also specified that the guards were to provide necessities to the householders.

On the subject of shutting up plague victims within their houses, Kellwaye seems to be considerably at odds with the law, at least as stated within London.

Kellwaye says "that when the infection is but, in few places," the sick should be restricted to their houses. This statement implicitly suggests that once the plague is distributed widely, Kellwaye did not think it necessary to restrict the movements of the sick. It is unclear whether Kellwaye believed that there was no point restricting the ill once plague was widely distributed, or if he was simply addressing the reality that during an intense epidemic, sick people restricted to their houses could not be properly cared for. Nathaniel Lodge also wrote that "many who were lost might have now been alive, had not the tragical mark upon their door drove proper assistance from them."

Despite the common belief that restrictions placed on assisting the ill were inflicting greater pain and suffering, a better solution was not found.

Unfortunately, it is unclear how well these quarantine measures were enforced. The Privy Council occasionally sent complaints to the Mayor decrying the City's slack enforcement, which suggests that enforcement was sporadic. The need for the plague orders to mandate guards at the houses, despite subsequent difficulties raising the money to pay them, also suggests that people attempted to avoid the restrictions of household quarantines. Writing in 1722, Defoe, in *A Journal of the Plague Year,* describes people employing a wide variety of tricks to avoid being locked in their houses. While we do not know that the stories, he reports are true, any more than whether Pepys' story about passing a naked baby to waiting friends is true, these stories provide evidence that people assumed quarantines were breached. Defoe endorses the idea of contagion and strongly recommends that during plague epidemics, healthy people should limit their contact with the outside world. However, he describes quarantines as overly punitive because they locked the well in with the sick, and were virtually useless because people did everything, they could to avoid them.

Pepys' Diary, written during the 1665 plague epidemic, recounts several stories that support Defoe's interpretation of behavior and attitudes toward these household quarantines. Pepys' story of a quarantined family that passed their naked child out of the house so that it would have a chance of survival. This story suggests not only that people considered the quarantines life-threatening but also that they were willing to do almost anything to circumvent the restrictions. If people were willing to go to great lengths to avoid being shut up in their houses, it is very unlikely that these measures were effective controls on the plague. As Paul Slack suggests, it is more likely that the macro-level quarantines, which prohibited travel and the importation of goods, along with rats and their fleas, from plague-infested areas, had an impact on the spread of the plague.

In Conclusion

As soon as the plague arrived in Europe, people began to battle it with all the weapons at their disposal. Ultimately, these weapons included many regulations. In England, these regulations restricted a wide range of behaviors and can be described as early attempts at public health measures; they demanded that the sick and the healthy be kept separate. People who had been in contact with the sick, or with the material possessions of the ill, were presumed to be ill until they had survived a quarantine period. The weapons in the battle against the spread of plague also included a broad array of changes to rules governing social functions and burial customs. The regulations consistently encouraged cleanliness and they even recommended burning potentially infective materials of little value.

Less directly, changes produced by the plague such as the population decrease and the corresponding increase in wealth produced changes in the class structure and commercial activities. London's population growth was in part the result of these developments. Also, although refuse and its stench were undoubtedly a constant problem within London, the plague called attention to the problems of garbage and human wastes and thus encouraged regulations to improve the environment. The social changes that resulted from plague and the battle against it affected the subsequent development of plague in ways that still are not fully understood but that must have involved the movement of rats and fleas and their contact with people.

Conclusion

To Commande that all those which doe visite and attende the sick, as also all those which have the sicknes on them and doe walke abroad: that they doe carry some thing in their handes therey to be known from other people. And here I must advertise you of one thing more which I had almost forgotten (which is) that when the infection is but in a few places, there to keepe all the people in the houses, not suffering any one of`them to goe abroad, and so to provide, that all such necessaries as they shall neéde may beé brought unto them during the time of their visitation: and when it is staide, then to cause all the clothes, bedding and other such thing as were used about the sicke, to be all burnt although at the

> *charge of the rest of the in habitants, you buy them all newe, for feare least the danger which may ensue thereby doe put you to a greater charge and grief: all these foresaide things are most dangerous, and may cause a generall infection, to the destroying of a whole Cittie.*

The Second Pandemic ravaged England and produced high mortalities for almost 320 years. During the pandemic, epidemics continued to break out sporadically throughout the country, although over time they became less universal and gradually became less frequent. Despite reductions in frequency, individual epidemics continued to produce devastating mortalities; even late in the pandemic, London experienced an epidemic plague about every 20 years, although the full extent of plague mortality is still debated. These repeated outbreaks exerted intense pressure on English society, especially in London, and the high mortality and plague's virtually constant presence inexorably changed England. In 1348 when the plague first arrived, the English was used to dealing with famine, the dearth of material necessities, and high mortality, especially among the very young, but the rapidity and severity of the initial wave of the epidemic produced a new level of societal stress. Nonetheless, the plague seems to have been a catalyst that both facilitated and quickened social changes that had already begun when the plague arrived. Some of these changes, in conjunction with experiences gained from living with plague, improved people's chances of avoiding the disease.

The country-wide, high mortalities of the first waves of the pandemic reduced the English population by about half, and it was not until the early sixteenth century that England's population regained its former numbers. But even while England's population remained low, London continued to grow, and by the seventeenth century, London's population was increasing dramatically. As London grew, its central core experienced some growth in population density, but of necessity, the surrounding suburbs became integral components of greater London although they were not within the jurisdiction of the London corporation.

Along with these demographic changes, the pattern of London epidemics changed. In the epidemic of 1563 and presumably in the earlier ones, the heaviest plague mortalities were clustered within London's walls, and wealthy parishes experienced mortalities as high as those in the poor parishes. By the epidemic of 1593, only a few years after London and national plague orders had begun to be printed, this pattern was reversed. Plague mortalities were focused outside the walls and in the poorest parishes. Although the plague orders do not seem to have offered especially innovative countermeasures, they regularized and made official concerns and reactions that had been expressed for years. The pattern of changes in mortality distribution during the last 100 years of plague in London suggests that factors in addition to new regulations affected the plague's distribution. Because mortality levels during this period became lower in wealthy parishes, I argue that material wealth as well as changes in behavior were responsible for this transformation. Supporting my contention is how subsequent changes, which have kept humans and rats at a greater distance, have prevented the plague from returning to Britain with the ferocity it exhibited during the Second Pandemic.

Notwithstanding the many changes that plague brought, the broad outline of responses to plague, and advice for avoiding the disease were remarkably consistent throughout this period. The first plague tracts focused on determining the causes of plague to a degree that was not repeated in later treatises, but they also provided advice and suggestions that continued to be offered throughout the entire pandemic. People were encouraged to flee areas of infection, to avoid low swampy areas, and to close south-facing windows. From the first plague outbreaks, clothes, fabric, and other objects belonging to plague victims were associated with spreading plague. Thus, cleaning and perfuming or fumigating, and burning fabrics were consistently seen as a means of preventing the spread of the plague. Because these materials were so precious, the cloth items of plague victims were likely reused despite the dangers they were thought to present. However, attitudes changed throughout the pandemic as material wealth increased, and as the association of cloth with the spread of plague became firmly fixed in the public consciousness. In England, some of these suggestions became written into law in the late 1500s: houses were quarantined, public gatherings were forbidden during periods of high plague mortality, and infected houses and fabrics were required to be aired and fumigated before being reused. Despite the apparent consistency in recommended responses to the plague, it appears that throughout the pandemic the effectiveness of these responses improved. This suggests that their implementation changed over time.

The idea that changes in behavior and culture had an impact on the plague is predicated on the fact that the plague was caused by *Yersinia pestis*. Despite considerable disagreement about what disease produced the high mortalities of the Second Pandemic, both direct and circumstantial evidence supports the argument that *Y. pestis* was the causative agent for most of the plague epidemics. Having said that, it is important to acknowledge that the epidemics of the Second Pandemic differ significantly from those of the

Modern Pandemic. The first wave of 1347-8, especially, traveled quickly and infected more territory faster than has any plague epidemic since. Additionally, morbidity levels in the Second Pandemic were much higher than any recorded, so far, during the Modern Pandemic. Furthermore, it is curious that Europeans never wrote about large die-offs of rats. Notwithstanding these differences, in other respects, the disease responsible for the heavy mortalities of the Second Pandemic closely resembles disease known to be produced by *Y. pestis*. The plague came in two forms, one form with a lung component that was extremely fatal and a second form that produced swellings in the lymph nodes. Also, occasionally people were reported as having been well in the morning but dead by nightfall, or were reported to have died of fear; these descriptions of deaths resemble deaths due to the third or septicemic form of plague.

Most importantly, the complexity of the plague's infectious cycle made it very difficult for early modern Europeans to determine its method of dispersal. Because of its dependence on rats and fleas, the plague could appear in widely dispersed houses and in people who thought that they had had no contact with plague-infected people. The complexity of the plague's dispersal pattern explains why people did not come to a consensus about whether the plague was extremely contagious or not.

The remarkable consistency of responses to plague could be used to argue that people simply clung to traditional responses because they did not understand with what they were dealing nor did they know how to respond to the crisis or have any real expectation of improved results. This view is much too simplistic, however. From the first outbreaks, physicians and surgeons used a blend of traditional learning and their observations to write tracts that attempted to make sense of the disease. Within a broad area of consensus, changes in the tone and emphasis of plague writings indicate that observers, commentators, and the medical community were struggling to understand epidemics and to evaluate the means to combat them in the light of empirically gained knowledge as well as tradition.

Although the traditional Greco-Roman medical understanding of disease did not view illnesses as discrete entities with specific causative agents, but rather as a result of humoral imbalance, the earliest plague tracts describe the epidemic in ways that suggest that it had a single cause. Initially, the plague was widely seen as both the result of some sort of terrible miasma-producing event and as a sign of God's displeasure. For the most part, the authors of the medical treatises situated the cause of the plague in astrological and large-scale environmental catastrophes that had enveloped the world in miasmic vapors. After the first wave of the pandemic, as plague appeared sporadically in apparently random locations, the plague was seen as the result of all sorts of factors, from the general sinfulness of society to nasty smells, bad foods, and tainted water to an overabundance of fruits, and miasma. The earliest treatises provided very little advice on how to cure the plague, but they did provide an array of suggestions for avoiding it. Although many of their explanations about the causes of plague and its method of dispersal seem fanciful, the doctors were proceeding logically based on their world view.

Although the physicians' earliest plague treatises

expressed ideas that were to become the foundation for plague control measures, they provided little information on treating people infected with the plague. Initially, plague treatises lacked treatment advice because the physicians and surgeons had no experience dealing with plague and because the plague was seen as virtually a death sentence. The medical advice published, along with the plague orders and Kellwaye's manual, clearly shows that by the early seventeenth century, doctors assumed that some patients would survive the plague and that proper care and attention could effect a cure. The medical community's understanding of the plague was relatively consistent, although it was neither uniform nor unchanging. Many of the preventative measures seem to have been based on the theoretically-driven university tradition rather than on any experiential or empirical evidence. For example, suggestions included avoiding southerly winds and eating a balancing diet of food low in moist and melancholic foodstuffs, but the advice was not limited to a strictly humoral conception that eschewed the idea of human agency in the spread of the plague. Medical suggestions for dealing with the plague were quickly influenced by experience.

Furthermore, despite the medical theory that varying symptoms were the result of different constitutions, people attempted to differentiate between diseases that were and were not plague. Some advice included in the first treatises, such as Foligno's advice to bleed people heavily, was not repeated in later treatises while less aggressive advice, such as the recommendation to bleed people only before sores became visible, became widely accepted. In general, recommended treatments for the plague became less aggressive and more palliative. Although some of Kellwaye's recommended curative measures seem unpleasant or bizarre, and it is hard to believe that anyone would follow all of them, overall, they suggest that people believed that intensive care for plague victims improved their chances of survival.

With the arrival of the first epidemic, physicians, surgeons, and chroniclers recommended flight as the best means of escaping plague, but modern epidemiology suggests that it was a flight that quickly spread the disease across the continent when the plague first appeared. As Samuel Pepys' Diary shows, in 1665 flight remained a popular and successful means of avoiding the plague. Those who could afford to leave London did. People fled both before and during the height of the epidemic, yet unlike in earlier epidemics, their flight did not spread the plague-like wildfire in the regions around London. Why were the results of flight so different in 1348 than in 1665? The causes of this differential are not known but are probably the result of many factors. The rats had probably acquired some resistance so rat colonies were not wiped out by any stray infected fleas the fleeing humans carried with them. Further, if people cleaned, shook, and aired their clothes, as the plague orders recommended, they were less likely to transport fleas to their new locations. Also, although people continued to leave London after the epidemic was in full swing, because of increased information about where plague victims were people fled earlier in the cycle before they had become infected. Moreover, people left London riding in carriages and carts rather than on foot.

Despite medical theory, when the plague first arrived, people saw it as contagious and attempted to avoid contact with those infected. The earliest chronicles describe the disintegration of society by giving examples of how fear of plague drove people apart. These early chronicles make it clear that from the outset, people accepted the idea that avoiding sick people was an effective means of avoiding the plague.

Nonetheless, despite the early acceptance of a generalized concept of contagion, it was not until 1578 that quarantine and sequestration regulations became the law of the land in England. These measures were part of a lengthy European tradition that had only minimal success. The regulations had two parts. One was the quarantining of ships from regions known to be infected with the plague. To the extent that new plague outbreaks were imported, quarantining ships probably reduced the number of plague outbreaks.

However, evidence, such as that the place of origin of the Great 1665 London plague epidemic was in the western suburbs rather than along the docks, suggests that epidemics arose from endemic rat infestations rather than by continually reintroduced by ships. The other element of quarantine regulations was the sequestering of people and their families within their own homes. This plague counter-measure is unlikely to have reduced plague mortality because it subjected entire households to infected fleas. Additionally, there is some evidence that wealthy people were less likely than the poor to be subject to these regulations, so this counter-measure may have contributed to the altered mortality distribution.

England does not seem to have experienced the extremes of societal breakdown that occurred in Europe. Even during the initial wave of plague, the kind of wild behavior common on the mainland, represented by violent public religious events or ceremonies and attempts to scapegoat minority groups, did not take place. In England, the plague seems to have been linked to the idea of chaos. As a result, the control of the plague included tactics to restrict social and environmental disorders. People, behavior, and disruptive animals were controlled. Vagabonds, foreigners, and poor day laborers living in crowded tenements were often treated suspiciously and were accused of spreading plague, although not necessarily on purpose. It is impossible to gauge the effect London's animal control efforts had on plague because this would depend not only on the ratio of cats and dogs to rats killed but also on when in the epidemic cycle they were killed. Terms used by Kellwaye and seventeenth-century Londoners to describe the crowded, dank and airless tenements and back alleyways where the plague was thought to flourish are very similar to those used in the earliest plague treatises to describe sources of plague such as closed up wells and caverns. On the other hand, it seems likely that efforts to strictly limit the growth of tenements probably made the conditions in those that did exist worse, which might have served to increase the plague differential between poor and wealthy parishes.

The role that religion played in combating plague changed substantially throughout the pandemic. Prayer was always recommended as a personal means of preventing and avoiding plague, but by1665 public religious observance was no longer an important aspect of the battle against the plague. In the fourteenth century, as the first wave of the plague epidemic swept across England, the Bishops of Lincoln and York called for public processions and large public prayers. Later, King Henry VIII also called for large public religious processions but throughout the Second Pandemic, public religious activities became a less significant aspect of plague control. In part, this change reflects the growing lack of religious uniformity in the country, yet, notably, parish churches were central to local plague control efforts. It is likely that this change in the role of the church in responding to the plague also reflects a changed understanding of the role of the natural world concerning the plague. Parish officers collected and distributed money for poor relief chose the searchers and maintained death records, which were an integral aspect of plague control. Throughout the pandemic, people came to believe that human agency could affect the course of epidemics.

Because of the association of bad smells with the disease, urban filth was associated with the disease even during the initial plague outbreak; thus, cleaning London was a significant feature of battling plague. In response to the epidemic of 1349, King

Edward ordered London streets cleaned of the human filth that was littering them, and during the second epidemic in 1361, he ordered animal butchering to take place outside the walls of London.

Most of the plague regulations and counter-measures took during the last one hundred years of plague in London were not innovative, although they did tie the control of disruptive behaviors, vagabonds, and indigent people and providing money for the care of the destitute together in new ways. The plague orders merely codified and published recommendations made when the plague first arrived. Nonetheless, recorded decreases in plague mortalities within central London, especially in the wealthier parishes, provides evidence that people who followed the orders reduced their chances of falling victim to the plague. Though they did not understand the disease as a discrete entity, nor did they conceive of plague as spread by rats and fleas, people changed their behavior in a way that affected plague distribution. Londoners' responses to plague, which included flight along with fumigation and cleaning of the town, their fabrics, and their residences seem to have altered the environment enough to increase the distance between people and rats and their fleas. The result was a reduced incidence of plague in those areas of London where the rules were enforced and where people had the resources to observe them.

CPSIA information can be obtained
at www.ICGtesting.com
Printed in the USA
LVHW081018211220
674520LV00008B/136